Eliot Attridge, Matthew Priestley, Dorothy W

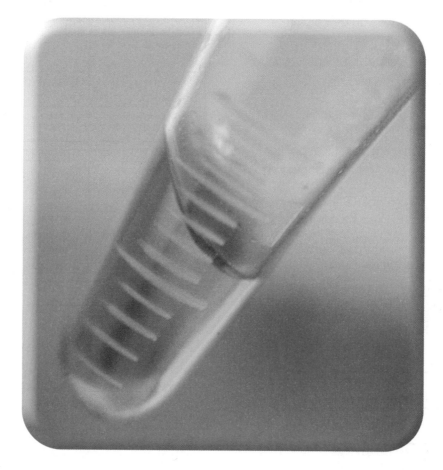

REVISION PLUS

OCR Twenty First Century

GCSE Additional Science A

Revision and Classroom Companion

Ideas about Science

Introduction to Ideas about Science

The OCR Twenty First Century Additional Science specification aims to ensure that you develop an understanding of science itself – of how scientific knowledge is obtained, the kinds of evidence and reasoning behind it, its strengths and limitations, and how far we can rely on it.

These issues are explored through Ideas about Science, which are built into the specification content and summarised over the following pages.

The tables below give an overview of the Ideas about Science that can be assessed in each unit and provide examples of content which support them in this guide.

Unit A162 (Modules B4, B5 and B6)

Ideas about Science	Example of Supporting Content
Data: their importance and limitations	Fieldwork (page 9)
Cause–effect explanations	Limitations to Photosynthesis (page 8)
Developing scientific explanations	Memory Models (page 29)
Making decisions about science and technology	Ethical Decisions (page 20)

Unit A172 (Modules C4, C5 and C6)

Ideas about Science	Example of Supporting Content
Data: their importance and limitations	Titration; Interpreting Results (pages 61–62)
Cause–effect explanations	Rates of Reactions; Measuring the Rate of Reaction; Analysing the Rate of Reaction; Changing the Rate of Reaction (pages 62–65)
Developing scientific explanations	The Development of the Periodic Table (page 32); Spectroscopy (page 34)
The scientific community	The Development of the Periodic Table (page 32)
Risk	Hazard Symbols; Safety Precautions; Group 1 – The Alkali Metals; Group 7 – The Halogens (pages 36–39)
Making decisions about science and technology	Metals and the Environment (page 51)

Unit A182 (Module P4, P5 and P6)

Ideas about Science	Example of Supporting Content
Data: their importance and limitations	Nuclear Fusion (page 89)
Cause–effect explanations	Collisions (page 76)
Developing scientific explanations	Nuclear Fusion (page 89)
Risk	Uses of Radiation (page 92)
Making decisions about science and technology	Tracers in the Body (page 93)

① Data: Their Importance and Limitations

Science is built on data. Scientists carry out experiments to collect and interpret data, seeing whether the data agree with their explanations. If the data do agree, then it means the current explanation is more likely to be correct. If not, then the explanation has to be changed.

Experiments aim to find out what the 'true' value of a quantity is. Quantities are affected by errors made when carrying out the experiment and random variation. This means that the measured value may be different to the true value. Scientists try to control all the factors that could cause this uncertainty.

Scientists always take repeat readings to try to make sure that they have accurately estimated the true value of a quantity. The mean is calculated and is the best estimate of what the true value of a quantity is. The more times an experiment is repeated, the greater the chance that a result near to the true value will fall within the mean.

The range, or spread, of data gives an indication of where the true value must lie. Sometimes a measurement will not be in the zone where the majority of readings fall. It may look like the result (called an 'outlier') is wrong – however, it does not automatically mean that it is. The outlier has to be checked by repeating the measurement of that quantity. If the result cannot be checked, then it should still be used.

Here is an example of an outlier in a set of data:

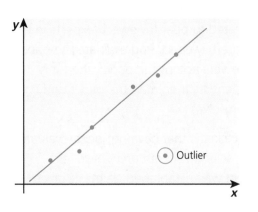

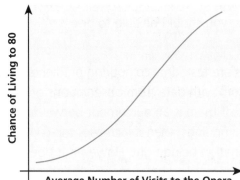

HT The spread of the data around the mean (the range) gives an idea of whether it really is different to the mean from another measurement. If the ranges for each mean do not overlap, then it is more likely that the two means are different. However, sometimes the ranges do overlap and there may be no significant difference between them.

The ranges also give an indication of reliability – a wide range makes it more difficult to say with certainty that the true value of a quantity has been measured. A small range suggests that the mean is closer to the true value.

If an outlier is discovered, you need to be able to defend your decision as to whether you keep it or discard it.

❷ Cause–effect Explanations

Science is based on the idea that a factor has an effect on an outcome. Scientists make predictions as to how the input variable will change the outcome variable. To make sure that only the input variable can affect the outcome, scientists try to control all the other variables that could potentially alter it. This is called 'fair testing'.

You need to be able to explain why it is necessary to control all the factors that might affect the outcome. This means suggesting how they could influence the outcome of the experiment.

A correlation is where there is an apparent link between a factor and an outcome. It may be that as the factor increases, the outcome increases as well. On the other hand, it may be that when the factor increases, the outcome decreases.

For example, there is a correlation between temperature and the rate of rusting – the higher the temperature, the faster the rate of rusting.

Just because there is a correlation does not necessarily mean that the factor causes the outcome. Further experiments are needed to establish this. It could be that another factor causes the outcome or that both the original factor and outcome are caused by something else.

The following graph suggests a correlation between going to the opera regularly and living longer. It is far more likely that if you have the money to go to the opera, you can afford a better diet and health care. Going to the opera is not the true cause of the correlation.

Sometimes the factor may alter the chance of an outcome occurring but does not guarantee it will lead to it. The statement 'the more time spent on a sun bed the greater the chance of developing skin cancer' is an example of this type of correlation, as some people will not develop skin cancer even if they do spend a lot of time on a sun bed.

To investigate claims that a factor increases the chance of an outcome, scientists have to study groups of people who either share as many factors as possible or are chosen randomly to try to ensure that all factors will present in people in the test group. The larger the experimental group, the more confident scientists can be about the conclusions made.

Ideas about Science

Even so, a correlation and cause will still not be accepted by scientists unless there is a scientific mechanism that can explain them.

3 Developing Scientific Explanations

Scientists devise hypotheses (predictions of what will happen in an experiment), along with an explanation (the scientific mechanism behind the hypotheses) and theories (that can be tested).

Explanations involve thinking creatively to work out why data have a particular pattern. Good scientific explanations account for most or all of the data already known. Sometimes they may explain a range of phenomena that were not previously thought to be linked. Explanations should enable predictions to be made about new situations or examples.

When deciding on which is the better of two explanations, you should be able to give reasons why.

Explanations are tested by comparing predictions based on them with data from observations or experiments. If there is an agreement between the experimental findings, then it increases the chance of the explanation being right. However, it does not prove it is correct. Likewise, if the prediction and observation indicate that one or the other is wrong, it decreases the confidence in the explanation on which the prediction is based.

4 The Scientific Community

Once a scientist has carried out enough experiments to back up his/her claims, they have to be reported. This enables the scientific community to carefully check the claims, something which is required before they are accepted as scientific knowledge.

Scientists attend conferences where they share their findings and sound out new ideas and explanations. This can lead to scientists revisiting their work or developing links with other laboratories to improve it.

The next step is writing a formal scientific paper and submitting it to a journal in the relevant field. The paper is allocated to peer reviewers (experts in their field), who carefully check and evaluate the paper. If the peer reviewers accept the paper, then it is published. Scientists then read the paper and check the work themselves.

New scientific claims that have not been evaluated by the whole scientific community have less credibility than well-established claims. It takes time for other scientists to gather enough evidence that a theory is sound. If the results cannot be repeated or replicated by themselves or others, then scientists will be sceptical about the new claims.

If the explanations cannot be arrived at from the available data, then it is fair and reasonable for different scientists to come up with alternative explanations. These will be based on the background and experience of the scientists. It is through further experimentation that the best explanation will be chosen.

This means that the current explanation has the greatest support. New data are not enough to topple it. Only when the new data are sufficiently repeated and checked will the original explanation be changed.

You need to be able to suggest reasons why an accepted explanation will not be given up immediately when new data, which appear to conflict with it, have been published.

5 Risk

Everything we do (or not do) carries risk. Nothing is completely risk-free. New technologies and processes based on scientific advances often introduce new risks.

Risk is sometimes calculated by measuring the chance of something occurring in a large sample over a given period of time (calculated risk). This enables people to take informed decisions about whether the risk is worth taking. In order to decide, you have to balance the benefit (to individuals or groups) with the consequences of what could happen.

For example, deciding whether or not to add chlorine to drinking water involves weighing up the benefit (of reducing the spread of cholera) against the risk of a toxic chlorine leak at the purification plant.

Risk which is associated with something that someone has chosen to do is easier to accept than risk which has been imposed on them.

> **HT** Perception of risk changes depending on our personal experience (perceived risk). Familiar risks (e.g. using bleach without wearing gloves) tend to be under-estimated, whilst unfamiliar risks (e.g. making chlorine in the laboratory) and invisible or long-term risks (e.g. cleaning up mercury from a broken thermometer) tend to be over-estimated.
>
> For example, many people under-estimate the risk that adding limescale remover and bleach to a toilet at the same time might produce toxic chlorine gas.

Governments and public bodies try to assess risk and create policy on what is and what is not acceptable. This can be controversial, especially when the people who benefit most are not the ones at risk.

6 Making Decisions about Science and Technology

Science has helped to create new technologies that have improved the world, benefiting millions of people. However, there can be unintended consequences of new technologies, even many decades after they were first introduced. These could be related to the impact on the environment or to the quality of life.

When introducing new technologies, the potential benefits must be weighed up against the risks.

Sometimes unintended consequences affecting the environment can be identified. By applying the scientific method (making hypotheses, explanations and carrying out experiments), scientists can devise new ways of putting right the impact. Devising life cycle assessments helps scientists to try to minimise unintended consequences and ensure sustainability.

Some areas of science could have a high potential risk to individuals or groups if they go wrong or if they are abused. In these areas the Government ensures that regulations are in place.

The scientific approach covers anything where data can be collected and used to test a hypothesis. It cannot be used when evidence cannot be collected (e.g. it cannot test beliefs or values).

Just because something can be done does not mean that it should be done. Some areas of scientific research or the technologies resulting from them have ethical issues associated with them. This means that not all people will necessarily agree with it.

Ethical decisions have to be made, taking into account the views of everyone involved, whilst balancing the benefits and risks.

It is impossible to please everybody, so decisions are often made on the basis of which outcome will benefit most people. Within a culture there will also be some actions that are always right or wrong, no matter what the circumstances are.

Contents

Contents

Module B4 (The Processes of Life)

Many different processes take place inside every cell of a living organism. This module looks at:

- the provision of molecules for energy and carbon skeletons via photosynthesis
- respiration under aerobic and anaerobic conditions
- enzymes and how they work
- the processes by which molecules enter and leave cells
- how to carry out sampling in fieldwork.

Chemical Reactions in Living Things

All living things are made from basic units. These are called **cells**.

Chemical reactions take place inside cells to enable them to function. All chemical reactions need energy to take place. The energy comes from the process of **respiration** – the release of energy from food.

The energy in food ultimately comes from the Sun. Plants (and some microorganisms, such as phytoplankton) at the start of food chains capture light energy by the process of photosynthesis, making food molecules. Energy is then transferred through food chains.

Photosynthesis

Photosynthesis can be summarised by saying that carbon dioxide and water are combined to produce sugar and oxygen in the presence of light and chlorophyll.

This sounds simple enough but in reality it just tells you the reactants needed at the start and the products created at the end. In reality photosynthesis is a series of reactions, some taking place in the light and some in the dark.

The sugar produced by photosynthesis is used as a carbon skeleton to build more organic molecules. All life is built from carbon skeletons in this way.

Respiration

Respiration can be summarised by saying that sugar and oxygen are combined to produce carbon dioxide and water, releasing energy in the process.

Respiration can be thought of as slow combustion. The sugar is the fuel.

As with photosynthesis, the summary above just tells you the reactants needed at the start and the products created at the end. Respiration is, in fact, a series of reactions releasing energy from large food molecules in all living cells.

Enzymes

Enzymes are organic catalysts. They are protein molecules that speed up the rate of chemical reactions in living organisms.

Enzymes are a type of protein coded for by instructions carried in genes.

The Lock and Key Model

Only a molecule with the correct shape can fit into an enzyme. This is a bit like a key (i.e. the molecule, called a substrate) fitting into the lock (i.e. the enzyme). Once the enzyme and molecule are linked the reaction takes place, the products are released and the process is able to start again.

The Lock and Key Model

Enzyme

Molecule

Combined molecule and enzyme. Reaction can take place.

Enzyme

Products

Molecule is broken down and enzyme can be reused.

Enzymes and Temperature

For enzymes to work to their optimum they need a specific and constant temperature. The graph below shows the effect of temperature on enzyme activity.

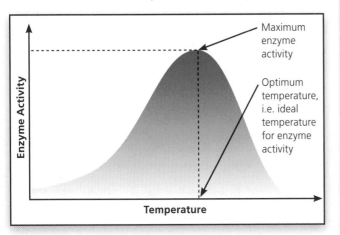

Different enzymes have different optimum working temperatures. For example, in the human body enzymes work best at about 37°C. Below this temperature their rate of action slows down while above 40°C they permanently stop working.

HT The enzyme activity at different temperatures is a balance between the increase in rate of reaction as the temperature increases and the changes to the active site at higher temperatures.

| Enzyme | | Heat (over 40°C) | | Enzyme permanently destroyed by heat |

Enzymes also have an optimum pH. For example, salivary amylase (an enzyme that breaks down starch in the mouth) works best at pH 6.8 (slightly acidic). Meanwhile, pepsin (an enzyme that breaks down protein in the stomach) works best at pH 1.5 (very acidic).

Enzymes can speed up catabolic reactions (where molecules are broken down) as well as anabolic reactions (where new molecules are built up).

HT Denaturing

The biological name for the process of permanent change in an enzyme's shape is **denaturing**. The enzyme becomes denatured.

At low temperatures, a small rise in temperature causes an increase in the rate of reaction.

The enzyme activity increases until the optimum temperature is reached. After this, the enzymes will be denatured – the active site (see below) will now be a different shape and the substrate will no longer fit.

A good analogy is a beef burger, made of protein. Before cooking, the burger is a particular shape. After cooking at high temperature, its shape is irreversibly changed.

The Active Site

The place where the substrate fits to the enzyme is called the **active site**. Each enzyme has a specific active site – only certain substrates will fit with it.

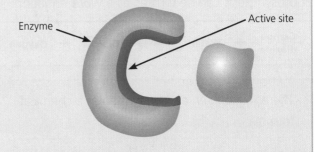

When subjected to high temperature, the shape of the active site **changes irreversibly**. This means molecules can no longer fit and the reaction stops. The enzyme is denatured.

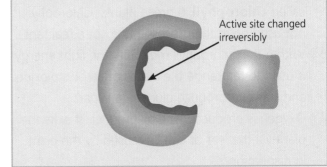

pH Changes

The active site is influenced by pH. Changes in pH in the enzyme's environment can make and break intra- and intermolecular bonds in the enzyme. This changes the shape of the active site and its effectiveness.

The graph below shows how changes in pH levels affect enzyme activity.

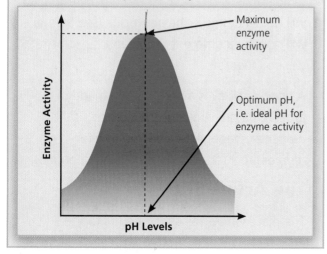

Maximum enzyme activity

Optimum pH, i.e. ideal pH for enzyme activity

Enzyme Activity

pH Levels

Photosynthesis

Photosynthesis can be written as a word equation:

Carbon dioxide + Water →(Light energy) Glucose [sugar] + Oxygen

The equation can also be written as a chemical formula:

$$6CO_2 + 6H_2O \xrightarrow{\text{Light energy}} C_6H_{12}O_6 + 6O_2$$

There are three stages in photosynthesis:

1. Light energy is absorbed by a green chemical called **chlorophyll** in green plants. Chlorophyll is not used up in the process – it is not a reactant.

2. Within the chlorophyll molecule, the light energy is used to rearrange the atoms of carbon dioxide and sugar to produce glucose (a sugar).

3. Oxygen is produced as a by-product. It exits the plant via the leaf or can be reused by the plant in respiration.

Glucose in Plants

Glucose is effectively a carbon skeleton that can be used to build many other molecules in living things. For example, cellulose is a structural carbohydrate that makes cell walls in plants. It is very strong and helps to provide support. Cellulose is made of repeating molecules of a certain form of glucose.

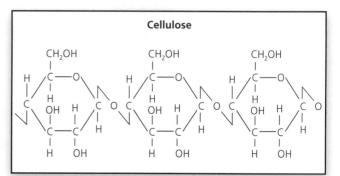

Cellulose

Protein is made up of different amino acid molecules, which themselves were made from glucose. They are vital for growth and repair.

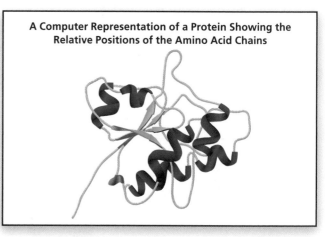

A Computer Representation of a Protein Showing the Relative Positions of the Amino Acid Chains

Chlorophyll is the organic pigment where the first stages of photosynthesis take place. It is also produced using a skeleton based on carbon.

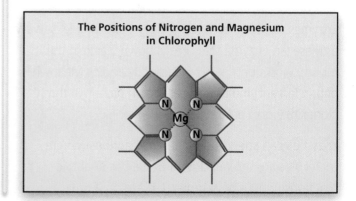

The Positions of Nitrogen and Magnesium in Chlorophyll

Storing Glucose

Glucose is converted into another carbohydrate suitable for storage, called **starch**. Starch is a long chain of glucose units that can be packed together very efficiently.

Glucose is used for energy, which is released by respiration.

Plant Cell Structure

The diagram below shows a plant cell:

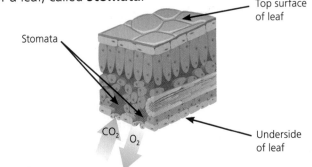

1. Chloroplasts – contain chlorophyll and the enzymes for some of the reactions of photosynthesis. They are only present in green parts of the plant
2. Cell membrane – allows dissolved gases and water to enter and leave the cell freely while acting as a barrier to other, larger chemicals
3. Nucleus – contains DNA that carries the genetic code for making enzymes and other proteins
4. Cytoplasm – where enzymes and other proteins are made
5. Mitochondria – where respiration occurs
6. Vacuole – used by the cell to store waste materials and to regulate water levels
7. Cell wall – provides support for the cell

Plant Transport

Plants need other chemicals in addition to glucose. The roots take up minerals from the soil in solution. Nitrogen, in the form of nitrates, is absorbed and used by the plant cells to make new proteins.

Substances move through cells via the process of **diffusion**. Diffusion is the overall movement of a substance from a region where it is in high concentration to an area where it is in lower concentration.

Diffusion is a passive process as it does not need an energy input to happen. For example, if the concentration of carbon dioxide in a plant cell is lower than the concentration outside, then carbon dioxide will diffuse into the cell.

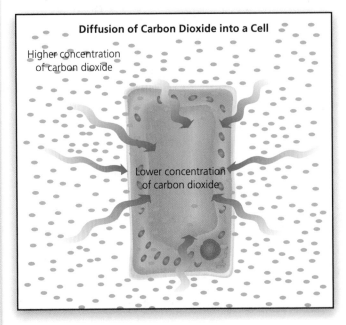

Diffusion of Carbon Dioxide into a Cell

Higher concentration of carbon dioxide

Lower concentration of carbon dioxide

Diffusion is the main method by which gases enter and leave the plant. Gases such as carbon dioxide and oxygen exchange between the leaf and the surrounding air through small holes on the underside of a leaf, called **stomata**.

Top surface of leaf

Stomata

CO_2 O_2

Underside of leaf

Osmosis

Osmosis is a specific type of diffusion. It is the overall movement of water from a dilute (i.e. with a high water to solute ratio) to a more concentrated solution (i.e. with a low water to solute ratio) through a **partially permeable** membrane.

A partially permeable membrane allows water molecules through, but not solute molecules because they are too large.

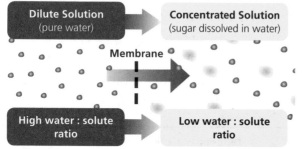

The effect of osmosis is to gradually dilute the concentrated solution. This is what happens at root hair cells, where water moves from the soil into the cells by osmosis due to a concentration gradient.

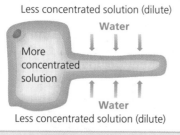

HT Active Transport

Active transport is the overall movement of a chemical substance against a concentration gradient (i.e. from where the substance is in low concentration to where it is in a higher concentration). This requires energy, which is provided by respiration.

An example of active transport is found in plant roots. Plants require nitrates. The nitrate concentration inside the plant is naturally higher than the outside. The plant cells use active transport to bring the nitrate from where it is in low concentration to where it is at a higher concentration in the root cell.

Limitations to Photosynthesis

There are several factors that can interact to limit the rate of photosynthesis:

- **Temperature** – too low and photosynthesis stops until the temperature rises again. Too hot and the enzymes stop working permanently.
- **Carbon dioxide concentration** – as carbon dioxide concentration increases, so does the rate of photosynthesis.
- **Light intensity** – light is needed for photosynthesis. The greater the availability of light, the quicker photosynthesis will take place.

You need to be able to interpret data on limiting factors for photosynthesis.

The graphs below each show a limiting factor (temperature, carbon dioxide or light intensity). The initial stage ① shows that the rate of photosynthesis increases.

At stage ② of each graph, the rate of reaction stays the same or, in the case of temperature, drops. This indicates that the rate of photosynthesis is now limited by one of the other factors (except for temperature, where enzymes are denatured so that photosynthesis stops).

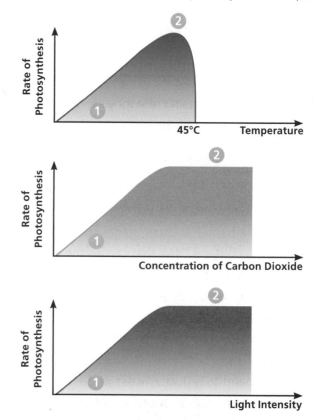

Fieldwork

To identify the effect of light on plants, biologists have to carry out fieldwork. This involves using a variety of techniques to measure the amount of available light and to see how this has affected the growth of plants.

A **light meter** can measure the amount of light that is hitting a leaf. The amount of light is measured in units of lux. Data-loggers can be fitted with a light meter and readings taken over a period of time.

It is impractical, if not impossible, to count all the plants growing in a given area. Instead, biologists sample a proportion of the available land to give an accurate estimate of plant cover. To determine this, biologists use **quadrats**. A quadrat is a square shape, often divided up into smaller squares. The quadrat is placed randomly on sections of the area in question and the plants that fall inside the area of the quadrat are counted.

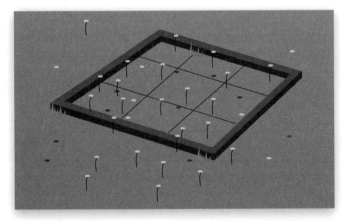

Alternatively, the area covered by the leaves of the plants can be counted. Sometimes a scale is used to indicate relative abundance, e.g. VC, C, UC, R, VR (very common, common, uncommon, rare, very rare).

It is vital to use a **key** to ensure that plants are correctly identified. A key enables the rapid identification of plants and animals by asking questions such as 'does the plant have parallel veins in its leaves?'

The answers rapidly lead to an idea of what the plant or animal is.

Example of a Key

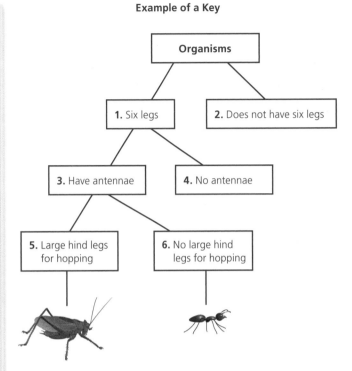

Sometimes it is preferable to measure the changes in plant life along a straight line, e.g. to see the succession of plants from a forest to the sea. In this case, a transect may be taken. A line is drawn and the quadrat is placed at set intervals along the line and the plants counted as before. This gives a picture of the changes in plant life over the line of the transect. In the example below, three fixed area plots are also investigated using randomised quadrats.

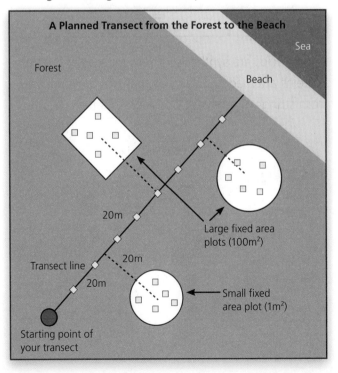

A Planned Transect from the Forest to the Beach

Respiration

All living organisms require energy for chemical reactions inside cells. This energy is released through the process of respiration. The energy is used for:

- movement
- synthesising (making) larger molecules
- **HT** active transport.

Synthesis of Large Molecules

Larger molecules, such as starch and cellulose, are synthesised from smaller molecules, such as glucose in plant cells. This involves joining the glucose molecules (monomers) together to form a polymer (made of many units).

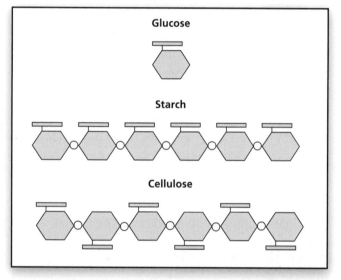

Amino acids are synthesised from glucose and nitrates. Proteins are made in plant, animal and bacterial cells from strings of amino acids joined together.

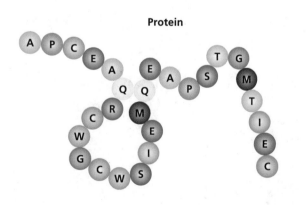

Aerobic Respiration

Aerobic respiration releases energy through the breakdown of glucose molecules, by combining them with oxygen inside living cells. The majority of animal and plant cells, and some microorganisms, respire aerobically.

The word equation for aerobic respiration is:

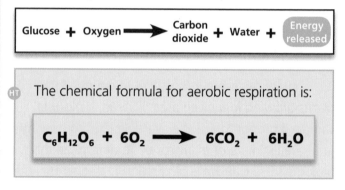

HT The chemical formula for aerobic respiration is:

$$C_6H_{12}O_6 + 6O_2 \longrightarrow 6CO_2 + 6H_2O$$

Anaerobic Respiration

Anaerobic respiration takes place in animal cells (e.g. in humans during vigorous exercise), plant cells (e.g. in plant roots in waterlogged soil) and in some microbial cells (e.g. bacteria in puncture wounds) when there is no oxygen or oxygen supply is low.

The equation for anaerobic respiration in animal cells and some bacteria is:

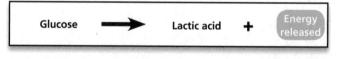

The equation for anaerobic respiration in plant cells and some microorganisms (including yeast, which is used in brewing and making bread) is:

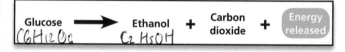

Aerobic respiration releases more energy per molecule than anaerobic respiration – a maximum of 18 times as much. In humans, anaerobic respiration can only occur for a short period.

Animal Cells

Animal cells have the following features:

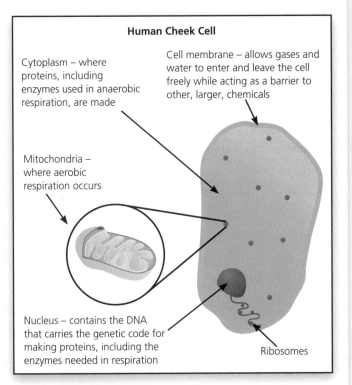

Human Cheek Cell

Cytoplasm – where proteins, including enzymes used in anaerobic respiration, are made

Cell membrane – allows gases and water to enter and leave the cell freely while acting as a barrier to other, larger, chemicals

Mitochondria – where aerobic respiration occurs

Nucleus – contains the DNA that carries the genetic code for making proteins, including the enzymes needed in respiration

Ribosomes

Microbial Cells

Microbial cells have similar structures, but with some important differences.

Bacteria

An important feature of bacteria is that they do not have any membrane-bound organelles. Therefore they do not have a nucleus or mitochondria:

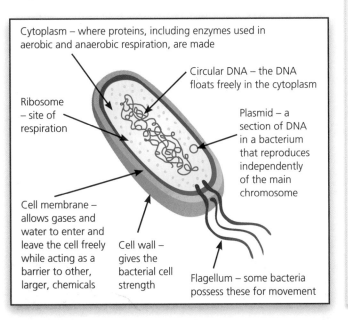

Cytoplasm – where proteins, including enzymes used in aerobic and anaerobic respiration, are made

Ribosome – site of respiration

Circular DNA – the DNA floats freely in the cytoplasm

Plasmid – a section of DNA in a bacterium that reproduces independently of the main chromosome

Cell membrane – allows gases and water to enter and leave the cell freely while acting as a barrier to other, larger, chemicals

Cell wall – gives the bacterial cell strength

Flagellum – some bacteria possess these for movement

Yeast

Yeast is a type of fungus. It is used to make bread and alcohol. Unlike bacteria, yeast does have membrane-bound organelles.

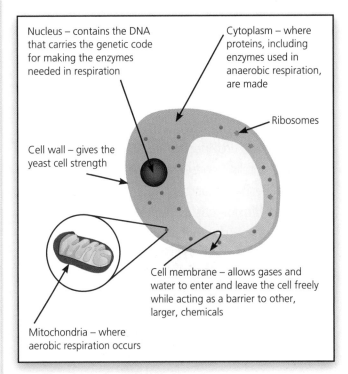

Nucleus – contains the DNA that carries the genetic code for making the enzymes needed in respiration

Cytoplasm – where proteins, including enzymes used in anaerobic respiration, are made

Ribosomes

Cell wall – gives the yeast cell strength

Cell membrane – allows gases and water to enter and leave the cell freely while acting as a barrier to other, larger, chemicals

Mitochondria – where aerobic respiration occurs

Cells Compared

The following table gives a comparison between the features of animal cells, bacteria and yeast:

	Animal Cell	Bacteria	Yeast
Cytoplasm	✓	✓	✓
Cell membrane	✓	✓	✓
Mitochondria	✓		✓
Nucleus	✓		✓
Ribosomes	✓	✓	✓
Cell wall		✓	✓
Circular DNA		✓	

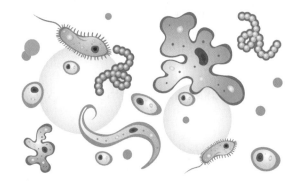

Applications of Anaerobic Respiration

Biotechnology has enabled us to use the products of anaerobic respiration.

Making Bread

Yeast is added to a dough, made from flour, salt, water and other ingredients. The dough is effectively a source of glucose that is needed for anaerobic respiration.

The dough also provides an anaerobic environment for the yeast cells to grow and multiply. As they grow, they respire anaerobically – producing ethanol and carbon dioxide gas as waste products.

The gas bubbles of carbon dioxide released by the yeast are trapped in a molecule called gluten. This makes the dough expand. The ethanol quickly evaporates.

Once the dough has risen and has been baked, all that is left is the expanded structure of the bread.

Brewing Alcohol

Brewing involves a fermentation process. There are two stages:

1 **Aerobic fermentation**

Aerobic fermentation lasts about a week and is the stage where the yeast is exposed to air and grows very rapidly on the sugar provided. Some alcohol is produced but the majority of energy is used to produce more yeast cells.

2 **Anaerobic fermentation**

Anaerobic fermentation lasts for weeks or months. This takes place in the absence of oxygen. The yeast respires anaerobically and produces alcohol and carbon dioxide instead of multiplying. Once brewed, the carbon dioxide escapes (unless it is needed by the brewer, e.g. in producing champagne).

Approximately 70% of the fermentation takes place aerobically. The remaining 30% of the fermentation takes place anaerobically. You might wonder whether you could get more alcohol if it was left longer. This would not work because the alcohol as a waste product is poisonous to the yeast and eventually it reaches a level high enough to kill the yeast. For stronger alcoholic content, distillation has to be undertaken.

Biogas

It is now possible to introduce bacteria to biodegradable substances such as manure, sewage and household waste in landfill sites. The anaerobic digestion leads to the production of methane (an explosive gas) and carbon dioxide.

The methane can be used as a low-cost fuel. As the methane is generated from materials that were thrown away, it is regarded as being a renewable energy source. The United Nations consider the gas to be a key energy source for the future.

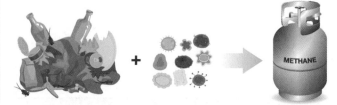

Module B5 (Growth and Development)

Genetic technologies, such as stem cell research and cellular-growth control, are at the cutting edge of modern science. This module looks at:

- how organisms produce new cells
- how genes control growth and development within the cell
- how new organisms develop from a single cell.

Cells, Tissues and Organs

Cells are the building blocks of all living things. Multicellular organisms are made up of collections of cells. The cells can become **specialised** to do a particular job.

Groups of specialised cells working together are called **tissues** and groups of tissues working together are called **organs**.

For example:

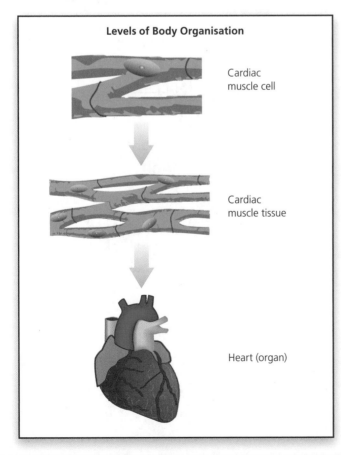

Levels of Body Organisation

Cardiac muscle cell

Cardiac muscle tissue

Heart (organ)

Mitosis

Mitosis is the process by which a cell divides to produce two new cells with identical sets of chromosomes to the parent cell. The new cells will also have all the necessary organelles.

The purpose of mitosis is to produce new cells for growth and repair and to replace old tissues.

Fertilisation

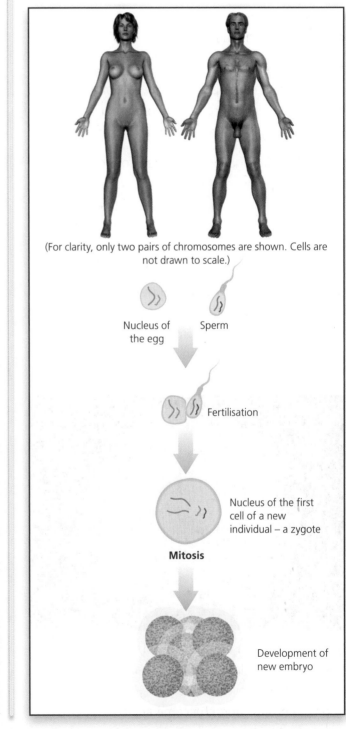

(For clarity, only two pairs of chromosomes are shown. Cells are not drawn to scale.)

Nucleus of the egg — Sperm

Fertilisation

Nucleus of the first cell of a new individual – a zygote

Mitosis

Development of new embryo

Fertilisation (Cont.)

When an egg is fertilised by a sperm it becomes a **zygote**.

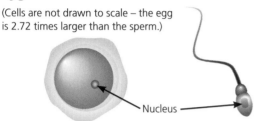

(Cells are not drawn to scale – the egg is 2.72 times larger than the sperm.)

Nucleus

The zygote then divides by mitosis to form a cluster of cells called an embryo.

Up to (and including) the eight cell stage, all the cells are identical and can produce **any** sort of cell required by the organism including neurons, blood cells, liver cells, etc. They are called **embryonic stem cells**.

At the 16 cell stage (approximately four days after fertilisation), most of the cells in an embryo begin to specialise and form different types of tissue. Although the cells contain the same genetic information, the position of each cell is different relative to the others. The distribution of various proteins in the cells will also be different. So, although the genes are the same, the cells are already subtly different from one another. At the time of specialisation these differences determine what specific functions a cell will have.

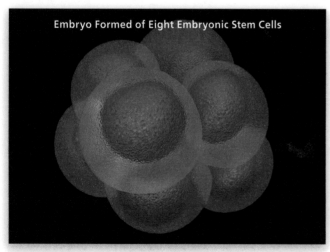

Embryo Formed of Eight Embryonic Stem Cells

Some cells remain unspecialised. These are **adult stem cells**. At a later stage they can become specialised. However, unlike embryonic stem cells, adult stem cells cannot become any type of cell.

Plant Meristems

Plants have cells that are like stem cells in animals. The cells are in areas called **meristems**. Only cells within meristems can divide repeatedly (i.e. are **mitotically active**).

Cells in the meristem are unspecialised but they can develop into any type of plant cell. Under normal hormonal conditions, this would mean that tissues such as xylem and phloem could be formed as well as organs such as leaves, roots and flowers.

There are two types of meristem: those that result in increased girth (**lateral meristems**) and those that result in increased height and longer roots (**apical meristems**):

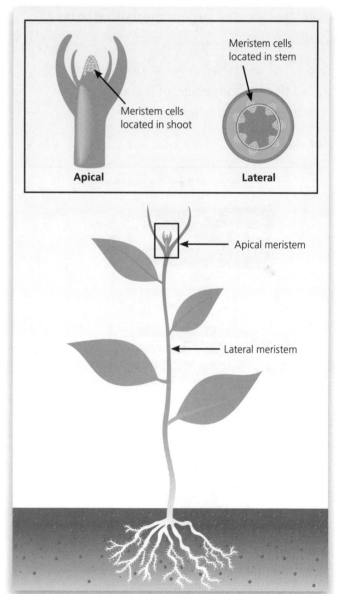

Meristem cells located in stem

Meristem cells located in shoot

Apical

Lateral

Apical meristem

Lateral meristem

Xylem and Phloem

Xylem is made from specialised cells to transport water and soluble mineral salts from the plant roots to the stem and leaves, and to replace water lost during transpiration and photosynthesis.

Phloem is made from specialised cells to transport dissolved food made by photosynthesis throughout the plant for respiration or storage.

When a stem is deliberately cut, special plant hormones can be added. These can send messages to the meristems to start to produce roots. As the cutting already has a stem and leaves, it will then grow into a clone of the parent plant.

> (HT) There are a wide range of plant hormones. The main group that is used in horticulture is called **auxin**. Auxins mainly affect cell division at the tip of a shoot, because that is where the meristems are. Just under the tip, the cells grow in the presence of auxins, causing the stem or root to grow longer.

The growth and development of plants are affected by the environment. One example is **phototropism**. Another example is geotropism. This is where roots and shoots grow towards and away from the source of gravity. Other tropisms also exist.

Phototropism

A tropism is the term given to a response by a plant to a stimulus. So, phototropism is a response by the plant to light.

A plant's survival depends on its ability to photosynthesise. Plants therefore need strategies to detect light and to respond to changes in intensity. This is demonstrated by the way in which plants will grow towards a light source.

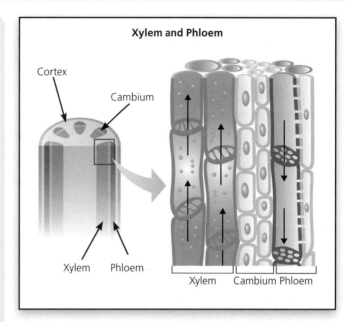

Xylem and Phloem

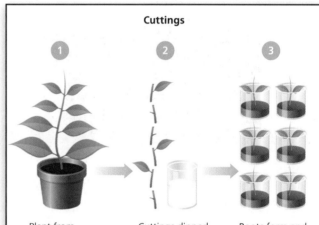

Cuttings

1. Plant from which cuttings are taken
2. Cuttings dipped into a rooting hormone
3. Roots form and new plants develop which are clones of the parent plant

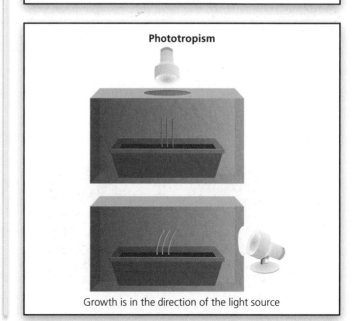

Phototropism

Growth is in the direction of the light source

How Phototropism Works

The cells furthest away from a light source grow more, due to the presence of auxin, which is sensitive to light.

Auxin is produced at the shoot tip and migrates down the shoot.

If a light source is directly overhead, then the distribution of auxin would be the same on both sides of the plant shoot.

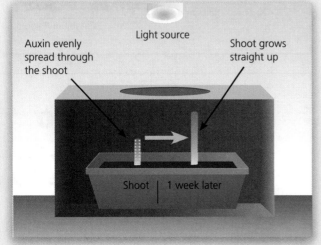

If a light source shines onto the shoot at an angle, the auxin facing the light moves to the side furthest away. The cells on the side furthest from the light source have proportionately more auxin than the closest.

As a result, the concentration of auxin on the side furthest away from the light increases, causing the cells there to elongate, and the shoot begins to bend towards the light.

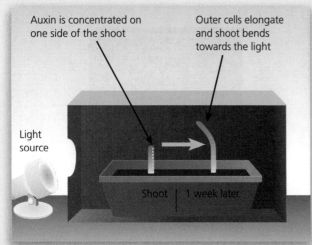

Charles Darwin carried out some simple experiments that demonstrated the role played by plant hormones produced in the shoot tip.

As we have seen, light causes the shoot to bend towards the source. If the tip of the shoot is removed or covered in opaque material then the plant will continue to grow upwards – as if the light source was not there.

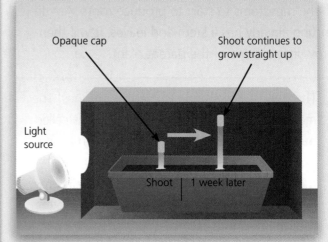

If the tip is covered with a transparent cap then it will still grow towards the light source. The same thing will happen if an opaque cylinder is wrapped around the stem leaving the tip exposed.

This experiment proves it is a substance produced in the tip that causes the cells further down the shoot to grow.

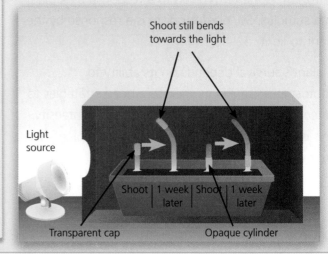

Mitosis and Growth

Mitosis leads to the production of two new cells, which are identical to each other and to the parent cell. Mitosis can only take place when a cell is ready to divide. This means that cells go through a **cell cycle**.

The cell cycle consists of a growth stage (G1) where the cell gets bigger and the number of organelles increase, then a synthesis stage (S) where the DNA is copied, followed by another very short growth stage (G2) immediately before mitosis (M phase). The cycle is a bit like a clock:

The Cell Cycle

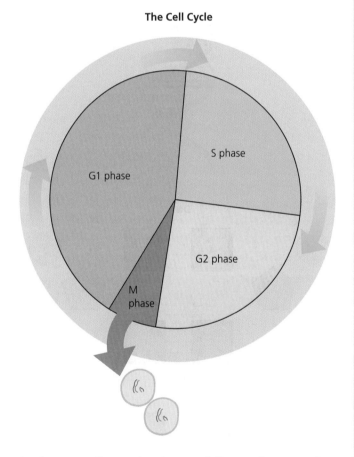

Both new cells need to have a full complement of organelles and DNA to function properly. Therefore, the number of organelles needs to increase and the DNA has to be copied.

The chromosomes are copied when the two strands of each DNA molecule separate and new strands form alongside them.

DNA Replication

The process of mitosis is shown in the following diagram:

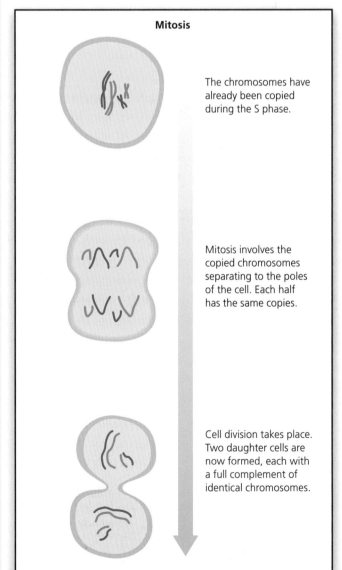

Mitosis

The chromosomes have already been copied during the S phase.

Mitosis involves the copied chromosomes separating to the poles of the cell. Each half has the same copies.

Cell division takes place. Two daughter cells are now formed, each with a full complement of identical chromosomes.

Meiosis

Meiosis only takes place in the testes and ovaries. It is a special type of cell division that produces gametes (sex cells, i.e. eggs and sperm) for sexual reproduction.

Gametes contain **half** the number of chromosomes of the parent cell. This is important because it means that when the male and female gametes fuse, the number of chromosomes will increase back to the full number. The resulting zygote has a set of chromosomes from each parent.

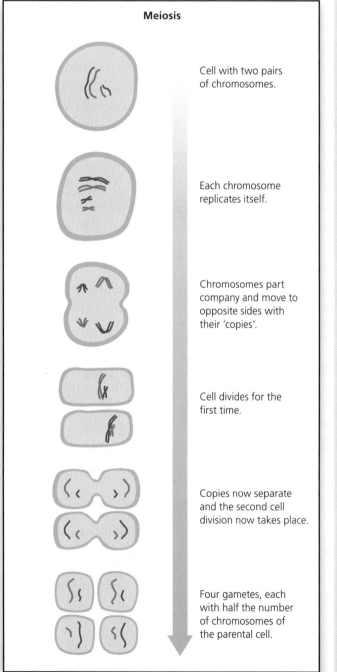

Meiosis

Cell with two pairs of chromosomes.

Each chromosome replicates itself.

Chromosomes part company and move to opposite sides with their 'copies'.

Cell divides for the first time.

Copies now separate and the second cell division now takes place.

Four gametes, each with half the number of chromosomes of the parental cell.

DNA

DNA is a nucleic acid, found in the nucleus, and is in the form of a double helix. The structure of DNA was worked out by Watson and Crick in 1952, earning them a Nobel prize for their discovery.

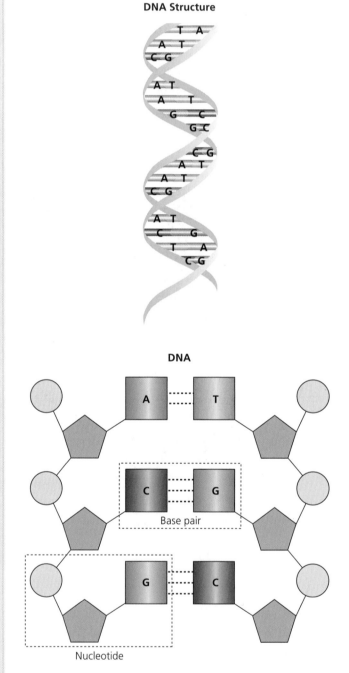

DNA Structure

DNA

Base pair

Nucleotide

The DNA molecule has a series of bases connected together, like rungs on a ladder. The bases are always in pairs:
- Adenine (A) pairs with thymine (T).
- Guanine (G) pairs with cytosine (C).

DNA Base Pairs

The DNA is always stored within the nucleus; it does not leave. The DNA has sequences of genes, which code for proteins. However, the proteins themselves are manufactured in the cytoplasm of the cell. Therefore, there is a mechanism for transferring the information stored in the genes into the cytoplasm.

Imagine the nucleus is a reference library, from which you are not allowed to remove the books. You would need to copy down any information you needed so that you could take it away with you.

The DNA molecule is too large to leave the cell. So, the relevant section of DNA is unzipped and the instructions copied onto smaller molecules, which can pass through the nuclear membrane of the nucleus into the cytoplasm.

HT The smaller molecules are called **messenger RNA** (mRNA). These leave the nucleus and carry the instructions to the **ribosomes**, which follow the instructions to make the specific protein.

It is the **order** of bases that determines what protein is made.

HT The bases in a gene are read in threes or triplets, called a codon. Each triplet means a specific amino acid is attached. As the amino acids join together a new protein is formed.

The mRNA is the opposite of the DNA code, with the exception that mRNA has the base uracil (U) instead of T – so, if the DNA base was A, the mRNA base would be U.

The triplet codes determine the amino acid produced. The table below helps to explain how. You do not need to recall it.

DNA Bases and their Complementary mRNA Bases

DNA Base	mRNA Base
A	U
T	A
G	C
C	G

Second Base of Codon

First Base of Codon	U	C	A	G	Third Base of Codon
U	UUU, UUC – Phenylalanine, Phe, (F); UUA, UUG – Leucine, Leu, (L)	UCU, UCC, UCA, UCG – Serine, Ser, (S)	UAU, UAC – Tyrosine, Tyr, (Y); UAA – Stop codon; UAG – Stop codon	UGU, UGC – Cysteine, Cys, (C); UGA – Stop codon; UGG – Tryptophan, Trp, (W)	U C A G
C	CUU, CUC, CUA, CUG – Leucine, Leu, (L)	CCU, CCC, CCA, CCG – Proline, Pro, (P)	CAU, CAC – Histidine, His, (H); CAA, CAG – Glutamine, Gln, (Q)	CGU, CGC, CGA, CGG – Arginine, Arg, (R)	U C A G
A	AUU, AUC, AUA – Isoleucine, Ile, (I); AUG – Methionine, Met, (M) Start codon	ACU, ACC, ACA, ACG – Threonine, Thr, (T)	AAU, AAC – Asparagine, Asn, (N); AAA, AAG – Lysine, Lys, (K)	AGU, AGC – Serine, Ser, (S); AGA, AGG – Arginine, Arg, (R)	U C A G
G	GUU, GUC, GUA, GUG – Valine, Val, (V)	GCU, GCC, GCA, GCG – Alanine, Ala, (A)	GAU, GAC – Aspartic acid, Asp, (D); GAA, GAG – Glutamic acid, Glu, (E)	GGU, GGC, GGA, GGG – Glycine, Gly, (G)	U C A G

Name of amino acid | Abbreviated name of amino acid | Single-letter code for amino acid

Start – when appearing for the first time, this causes the gene to be read.

Stop – when this codon is read it causes the transcription process to stop completely.

Gene Switching

It is a fundamental premise in biology that a gene codes for one specific protein. All body cells, including stem cells, contain **exactly the same genes**.

All cells have to produce proteins for growth and respiration. This means that certain genes (e.g. the gene for making the cell membrane and the gene for making ribosomes) will be switched on in all cells. The genes that are not needed will be switched off.

This can be visualised as:

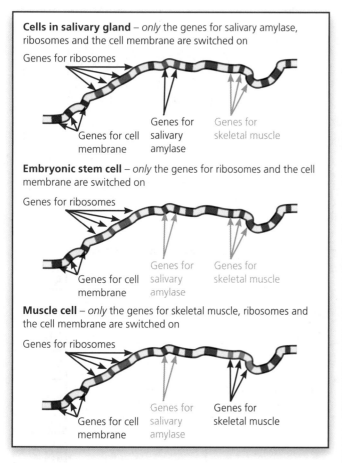

Cells in salivary gland – *only* the genes for salivary amylase, ribosomes and the cell membrane are switched on

Genes for ribosomes

Genes for cell membrane

Genes for salivary amylase

Genes for skeletal muscle

Embryonic stem cell – *only* the genes for ribosomes and the cell membrane are switched on

Genes for ribosomes

Genes for cell membrane

Genes for salivary amylase

Genes for skeletal muscle

Muscle cell – *only* the genes for skeletal muscle, ribosomes and the cell membrane are switched on

Genes for ribosomes

Genes for cell membrane

Genes for salivary amylase

Genes for skeletal muscle

When a stem cell becomes specialised, the genes for proteins specific to the new cell type will be switched on. *Any* gene in an embryonic stem cell *could* potentially be switched on, to make the specialised cell type needed.

In human DNA, there are approximately 20 000–25 000 genes, meaning a similar number of different proteins could be made. Each cell type will have a small fraction of these genes switched on.

Stem cells have the potential to produce cells needed to replace damaged tissues. For example, stem cells can be used to replace brain tissue in a patient with Parkinson's disease, or to grow new skin tissue following a burn.

Ethical Decisions

To produce the large number of stem cells needed for treatments, it is necessary to **clone** cells from five-day-old embryos. The stem cells are collected when the embryo is made up of approximately 150 cells. The rest of the embryo is destroyed. At the moment, unused embryos from IVF treatments are used for stem cell research.

There is an ethical issue as to whether it is right to use embryos to extract stem cells in this way. The debate revolves around whether or not the embryos should be classed as people.

One view is that if an embryo was left over from IVF (and would therefore never grow into a human being), it would be acceptable for stem cell research to be carried out on it as long as the parents gave their consent. However, another view is that destroying an embryo amounts to destroying a life.

The Government regulates and makes laws on such matters.

Mammalian Cloning

Whole animals have been cloned – Dolly the Sheep was the first to be cloned from adult skin cells. It is, however, illegal to clone a human being in this way.

Scientists can now take a mature, specialised cell and reactivate (switch on) inactive genes, effectively making it a new specialised cell type. This gives the potential to grow new tissue that is genetically the same as the patient. The benefit is that the tissue will not be rejected by the immune system of the patient. It also means that the patient does not have to take an expensive cocktail of drugs in order to prevent rejection. Such drugs can affect a patient's immune system and can stop it from fighting disease.

Module B6 (Brain and Mind)

Neuroscience (the study of the nervous system, including the brain) is an area at the forefront of medical research, as developments and discoveries have a huge potential impact for society. This module looks at:

- how organisms respond to changes in their environment
- what reflex actions are
- how information is passed through the nervous system
- how organisms develop more complex behaviour
- how we know about the way in which the brain coordinates the senses
- how drugs affect our nervous system.

Reflexes

Living organisms can detect and respond to a **stimulus**, i.e. a change in the environment, such as light, temperature, etc.

Receptors are stimulated by the stimulus and produce a rapid, involuntary (automatic) response. In other words, the organism responds without thinking. This is called a **simple reflex**.

The simplest animals rely on reflex actions for much of their behaviour. All their movement and reactions are simple reflex responses. The reflex actions ensure that the animal will respond in a way that is most likely to result in its survival.

For example, a simple reflex response to chemicals can lead to an organism finding food quickly. A change in light level could indicate the presence of a predator, so the organism moves away.

Simple Reflexes in Humans

Newborn babies cannot think for themselves – they exhibit a range of simple reflexes for a short time after birth, which ensures that they can survive. This is a little like having a start-up disk for a computer.

The absence of the reflexes, or their failure to disappear over time, may indicate that the nervous system of the baby is not developing properly.

Some simple reflexes exhibited by newborn babies include the following:

- **Stepping reflex** – when held under its arms in an upright position with its feet on a firm surface, a baby makes walking movements with its legs.
- **Grasping reflex** – a baby tightly grasps a finger that is put in its hand.
- **Sucking reflex** – a baby sucks on a finger (or the mother's nipple) when it is put into its mouth.
- **Startle reflex** – a baby shoots out its arms and legs when startled, e.g. by a sudden loud noise.
- **Rooting reflex** – a baby turns its head and opens its mouth, ready to feed, when its cheek is stroked.

Adults also exhibit a range of simple reflexes. They are the most efficient way of quickly responding to potentially dangerous events:

- **Pupil reflex** – bright light causes muscles in the iris in the eye to contract so that the retina is not damaged.

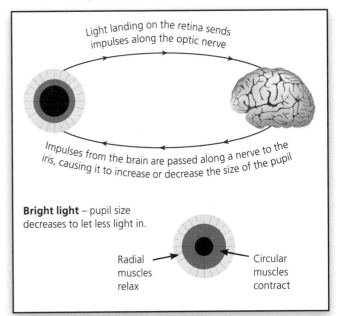

Light landing on the retina sends impulses along the optic nerve

Impulses from the brain are passed along a nerve to the iris, causing it to increase or decrease the size of the pupil

Bright light – pupil size decreases to let less light in.

Radial muscles relax

Circular muscles contract

- **Knee-jerk reflex** – when the knee is struck just below the knee cap, the leg will kick out.
- **Dropping hot object reflex** – when picking up a very hot object, the response is to throw it away in order to prevent heat damage to the hand.

Sending Signals

There are two ways of sending signals in the body. The first is via **electrical impulses** through long, wire-like cells called **neurons** (nerve cells). This method is very quick and short-lived.

Sensory neurons carry nervous impulses (electrical signals) from receptors to the central nervous system:

Impulse travels towards cell body

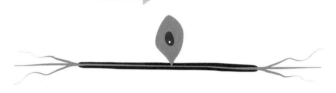

Motor neurons carry impulses from the central nervous system to effectors:

Impulse travels away from cell body

The other way signals are sent in the body is via chemicals called **hormones**, for example insulin (which controls blood sugar levels) and oestrogen (the female sex hormone), which are secreted into the blood. Chemical signals are slower than electrical impulses and they move to target organs, but their effect lasts a long time.

The nervous and hormonal communication systems are seen in larger, more complex organisms. This is a result of the **evolution** of multicellular organisms.

Detecting Changes

Nervous coordination in an animal requires the presence of one or more different **receptors** to detect stimuli. For example:

- **Light** – detected by receptors in the eyes
- **Sound** – detected by receptors in the ears
- **Changes of position** – detected by receptors for balance in the inner ear
- **Taste** – detected by receptors on the tongue
- **Smell** – detected by receptors in the nose
- **Pressure** – detected by receptors for pressure in the skin
- **Temperature** – detected by receptors for temperature in the skin

Coordinating the Response

The receptors are connected to a processing centre by sensory neurons. With simple reflexes, the processing centre is the spinal cord; the brain is not involved. The processing centre coordinates a response by sending back a message electrically via motor neurons to the **effector**, which carries out the response. This is called a **spinal reflex arc**.

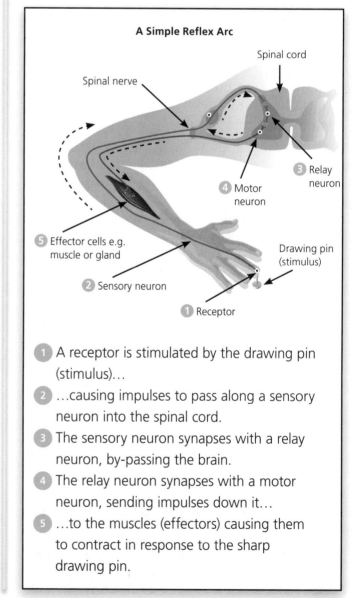

A Simple Reflex Arc

Spinal cord

Spinal nerve

3 Relay neuron

4 Motor neuron

5 Effector cells e.g. muscle or gland

Drawing pin (stimulus)

2 Sensory neuron

1 Receptor

1 A receptor is stimulated by the drawing pin (stimulus)…

2 …causing impulses to pass along a sensory neuron into the spinal cord.

3 The sensory neuron synapses with a relay neuron, by-passing the brain.

4 The relay neuron synapses with a motor neuron, sending impulses down it…

5 …to the muscles (effectors) causing them to contract in response to the sharp drawing pin.

The arrangement of neurons into a fixed pathway in a spinal reflex arc means that the responses are automatic and, hence, very rapid because no processing is required.

If the signal had to travel to the brain and be processed before action was taken, then, by the time the response arrived, it may be too late.

Receptors and Effectors

Receptors and effectors can form part of complex organs.

Muscle Cells in Muscle Tissue

The specialised cells that make up muscle tissues are effectors. Impulses travel along motor neurons and terminate at the muscle cells. These impulses cause the muscle cells to contract.

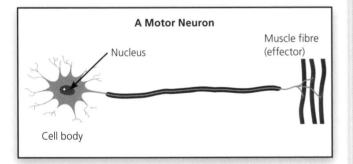

A Motor Neuron

Nucleus

Muscle fibre (effector)

Cell body

Light Receptors in the Retina of the Eye

The eye is a complex sense organ. The lens focuses light onto receptor cells in the retina, which are sensitive to light. The receptor cells are then stimulated and send electrical impulses along the sensory neurons to the brain.

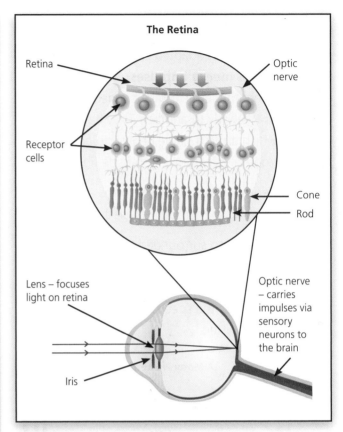

The Retina

Retina

Optic nerve

Receptor cells

Cone

Rod

Lens – focuses light on retina

Optic nerve – carries impulses via sensory neurons to the brain

Iris

Hormone Secreting Cells in a Gland

The hormone secreting cells in glands are effectors. They are activated by an impulse, which travels along a motor neuron from the **central nervous system** and terminates at the gland. The impulse triggers the release of the hormone into the bloodstream, which transports it to the sites where it is required.

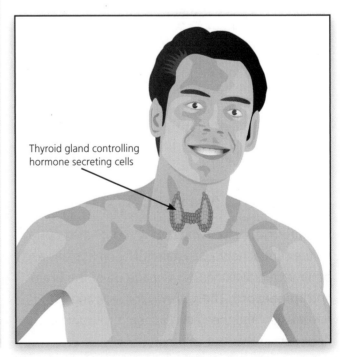

Thyroid gland controlling hormone secreting cells

The Structure of Neurons

Neurons are specially adapted cells that can carry electrical signals (**nerve impulses**). They are elongated (lengthened) to make connections from one part of the body to another. They have branched endings, which allow a single neuron to act on many other neurons or effectors, e.g. muscle fibres.

In motor neurons, the cytoplasm forms a long fibre surrounded by a cell membrane called an **axon**.

Some axons are also surrounded by a fatty sheath, which insulates the neuron from neighbouring cells (a bit like the plastic coating on a copper electrical wire) and increases the speed at which the nerve impulse is transmitted.

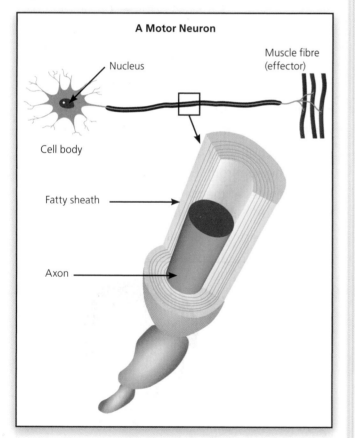

A Motor Neuron

Nucleus

Muscle fibre (effector)

Cell body

Fatty sheath

Axon

The Central Nervous System

The information from neurons is coordinated overall by the central nervous system (CNS). In humans and other vertebrates the CNS is made up of the **brain** and **spinal cord**. The pathway for receiving information and then acting upon it is shown in the flowchart in the next column.

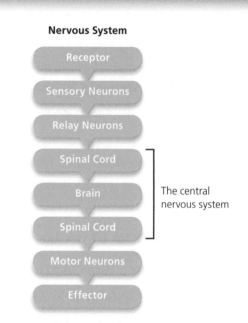

Nervous System

Receptor

Sensory Neurons

Relay Neurons

Spinal Cord

Brain — The central nervous system

Spinal Cord

Motor Neurons

Effector

The CNS is connected to the body via sensory and motor neurons, which make up the **peripheral nervous system** (**PNS**). The PNS is the second major division of the nervous system. Its sensory and motor neurons transmit messages all over the body, e.g. to the limbs and organs. They also transmit messages to and from the CNS.

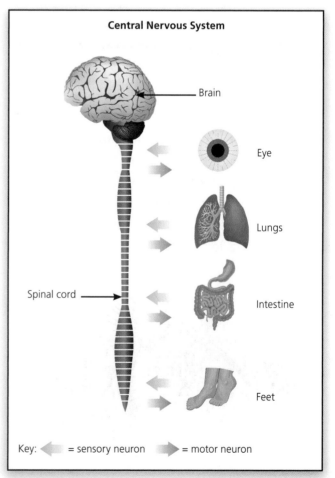

Central Nervous System

Brain

Eye

Lungs

Spinal cord

Intestine

Feet

Key: ◀ = sensory neuron   ▶ = motor neuron

Synapses

Synapses are the gaps between adjacent neurons. They allow the brain to form interconnected neural circuits. The human brain contains a huge number of synapses. There are approximately 1000 trillion in a young child. This number decreases with age, stabilising by adulthood. The estimated number of synapses for an adult human varies between 100 and 500 trillion.

HT When an impulse reaches the end of a sensory neuron, it triggers the release of chemicals, called **transmitter substances**, into the synapse. They diffuse across the synapse and then bind with specific receptor molecules on the membrane of a relay neuron.

The receptor molecules will only bind with specific chemicals to initiate a nerve impulse in the relay neuron, so the signal can continue on its way. Meanwhile, the transmitter substance is reabsorbed back into the sensory neuron, to be used again.

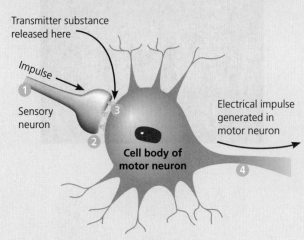

Transmitter substance released here

Impulse

Sensory neuron

Cell body of motor neuron

Electrical impulse generated in motor neuron

The sequence is as follows:
1. Electrical signal (nerve impulse) moves through sensory neuron.
2. Transmitter substances are released into the synapse.
3. Transmitter substances bind with receptors on the motor neuron.
4. Electrical signal (nerve impulse) is now sent through motor neuron.

Drugs and the Nervous System

Many drugs, such as Ecstasy, beta blockers and Prozac, cause changes in the speed at which nerve impulses travel to the brain, speeding them up or slowing them down. Sometimes false signals are sent.

Drugs and toxins can prevent impulses from travelling across synapses or they can cause the nervous system to become overloaded with too many impulses. For example, Ecstasy (an illegal recreational drug) and beta blockers (a drug used to prevent heart attacks) both affect the transmission of nerve impulses.

HT The drug Ecstasy, scientifically known as MDMA, in the nervous system affects a transmitter substance called **serotonin**. Serotonin can have mood-enhancing effects, i.e. it is associated with feeling happy.

Serotonin passes across the brain's synapses, landing on receptor molecules. Serotonin that is not on a receptor is absorbed back into the transmitting neuron by the transporter molecules. Ecstasy blocks the sites in the brain's synapses where the chemical serotonin is removed.

As a result, serotonin concentrations in the brain increase and the user experiences feelings of elation. However, the neurons are harmed in the process and a long-term consequence of taking Ecstasy can be memory loss.

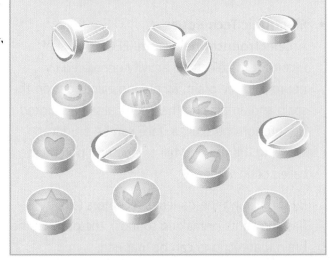

The Cerebral Cortex

The **cerebral cortex** is the part of the brain most concerned with intelligence, memory, language and consciousness (our being aware of our own thinking and existence). Scientists have used a variety of methods to map the different regions of the cerebral cortex:

- **Physiological Techniques**

  Damage to different parts of the brain can produce different problems, e.g. long- and short-term memory loss, paralysis in one or more parts of the body, speech loss, etc. Studying the effects of accidents or illnesses, as well as by directly stimulating the brain with electrical impulses, has led to an understanding of which parts of the brain control different functions:

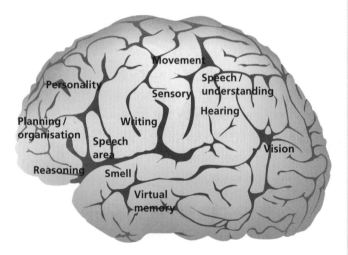

N.B. You do not need to learn the different parts of the brain.

- **Electronic Techniques**

  An **electroencephalogram** (EEG) is a visual record of the electrical activity generated by neurons in the brain. By placing electrodes on the scalp and amplifying the electrical signals picked up through the skull, a trace can be produced showing the rise and fall of electrical potentials called brain waves.

  By stimulating the patient's receptors (e.g. by flashing lights or making sounds), the parts of the brain that respond can be mapped.

Magnetic Resonance Imaging (MRI) scanning is a technique that produces images of cross-sections of the brain, showing its structure. The computer-generated picture uses colour to represent different levels of electrical activity. The activity in the brain changes depending on what the person is doing or thinking.

In 2010, patients in a persistent vegetative state after being involved in serious accidents were tested using MRI. It was shown that, although they were outwardly comatose, they could answer questions by imagining playing tennis to represent 'yes' and sitting down to represent 'no'. The parts of the brain that show increased electrical activity when playing or imagining playing tennis are different to those that show electrical activity when doing or thinking about something else.

Composite Image of MRI Scans of a Patient Showing Different Sections of the Head and Brain

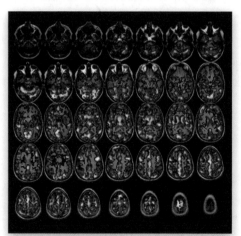

MRI Scanner

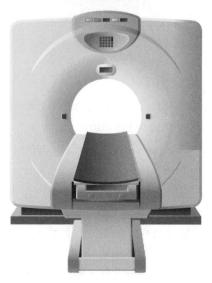

Conditioned Reflexes

Although they are not conscious actions, reflex responses to a new stimulus can be learnt. Through a process of conditioning, the body learns to produce a specific response when a certain stimulus is detected. Conditioning works by building an association between the new stimulus (the **secondary stimulus**) and the stimulus that naturally triggers the response (the **primary stimulus**). This resulting reflex is called a **conditioned reflex action**.

The effect was discovered at the beginning of the 20th century by a Russian scientist named Pavlov, who received a Nobel prize for his work.

Pavlov observed that whenever a dog sees and smells a piece of meat, it starts to salivate (produce saliva). In his experiment, a bell was rung repeatedly whenever meat was shown and given to the dog. Eventually, simply ringing the bell, without any meat present, caused the dog to salivate.

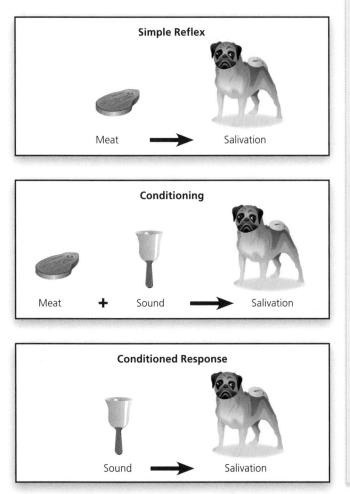

Simple Reflex

Meat ➡ Salivation

Conditioning

Meat **+** Sound ➡ Salivation

Conditioned Response

Sound ➡ Salivation

Another example of a conditioned reflex is when, after being stung by a wasp, you associate the yellow and black stripes with the painful sting. The next time you see a wasp (or similar insect such as a hoverfly), you feel fear.

HT In a conditioned reflex, the final response has no direct connection to the stimulus.

For example, the ringing bell in Pavlov's experiment is a stimulus that has *nothing* to do with feeding, it is just a sound. However, the association between the sound of the bell and the meat was strong enough to induce the dog to salivate.

Some conditioned reflexes can increase a species' chance of survival:

Example
The caterpillar of the cinnabar moth is black and orange in colour, to warn predators that it is poisonous. After eating a few cinnabar caterpillars, a bird will start to associate the colours with a very unpleasant taste and will avoid eating anything that is black and orange in colour (even if they are not actually poisonous).

In some situations, the brain can override or modify a reflex action.

For example, when the hand comes into contact with something hot, the body's natural reflex response is to pull away or drop the object.

However, you might know an object is hot but still want to pick it up, e.g. a hot potato or a dinner plate. To allow this, the brain sends a signal via a neuron to the motor neuron in the reflex arc, modifying the reflex so that you do not drop the potato or plate.

Development of the Brain

Mammals have a complex brain that contains billions of neurons. This enables them to learn from experience, including how to respond to different situations, e.g. social behaviour (dogs can be trained to be passive or aggressive).

The evolution of one particular mammal, *Homo sapiens*, led to the development of a larger brain than in other animals. Early humans could use tools, coordinate hunting and formulate plans about what might happen in the future. Having a larger brain meant that early humans were more likely to survive and reproduce, passing on the genes for producing a larger brain.

In mammals, neuron pathways are formed in the brain during development. The way in which the animal interacts with its environment determines what pathways are formed.

> **HT** It is the huge variety of pathways available that makes it possible for the animal to adapt to new situations.

During the first few years after birth, the brain grows very rapidly. As each neuron matures it sends out multiple branches, increasing the number of synapses.

At birth, in humans, the cerebral cortex has approximately 2500 synapses per neuron. By the time an infant is two or three years old, the number of synapses is approximately 15 000 synapses per neuron.

Each time an individual has a new experience, a different pathway between neurons is stimulated. Every time the experience is repeated, the pathway is strengthened. Pathways that are not used regularly are eventually deleted. Only the pathways that are activated most often are preserved.

These modifications mean that certain pathways of the brain become more likely to transmit impulses than others and the individual will become better at a given task. This is why certain tasks may be learned through repetition, e.g. riding a bike, revising for an exam, learning to ski or learning to play a musical instrument.

HT Feral Children

If neural pathways are not used then they are deleted. There is evidence to suggest that because of this, if a new skill (e.g. learning a language) has not been learned by a particular stage in development, an animal or child might not be able to learn it in the same way as normal.

One example of evidence showing this comes from the study of so-called feral children ('feral' means 'wild').

Feral children are children who have been isolated from society in some way, so they do not go through the normal development process. This can be deliberate (e.g. inhumanely keeping a child in a cellar or locked room), or it can be accidental (e.g. through being shipwrecked).

In the absence of any other humans, the children do not ever gain the ability to talk (or they lose any ability they had already gained) other than making rudimentary grunting noises. Learning a language later in life is a much harder and slower process.

Child Development

After children are born, there are a series of milestones that can be checked to see if development is following normal patterns.

If these milestones are missing or are late, it could mean that there are neurological problems or that the child is lacking stimulation (is not being exposed to the necessary experiences).

For example, at three months, babies should be able to lift their heads when held to someone's shoulder or grasp a rattle when it is given to them. By about 12 months, babies should be able to hold a cup and drink from it, and walk when one of their hands is held.

Memory

Memory is the ability to store and retrieve information. Scientists have produced models to try to explain how the brain facilitates this but, so far, none have been able to provide an adequate explanation. Current models are limited.

Verbal memory (words and labels) can be divided into **short-term** and **long-term memory**. Short-term memory is capable of storing a limited amount of information for a limited amount of time (roughly 15–30 seconds). Long-term memory can store a seemingly unlimited amount of information indefinitely.

When using short-term memory, it is currently thought that up to seven (+/- two) separate pieces of information can be stored (e.g. for a number with eight digits like 24042003, each number could fill a slot if eight were available).

This capacity can be increased by **chunking** the information, i.e. putting it into smaller chunks. For example, the number 201054738087 could be stored as (2010), (5473) and (8087), using only three of the seven units of storage.

Long-term memory is where information is stored in the brain through repetition, which strengthens and builds up neuron pathways.

Remember, Remember...

Humans are more likely to remember information when:

- it is repeated (especially over an extended period of time), e.g. going over key points several times as a method of revising for exams
- there is a strong stimulus associated with it, such as colour, light, smell or sound (the more senses that are involved, the better)
- there is a pattern to it (or if a pattern can be artificially imposed upon it). For example, people may find remembering the order of how organisms are classified difficult, so a sentence where there is a logical relationship between the words (unlike the names of the taxonomical groups themselves) can be remembered as a prompt instead: **K**ids **P**refer **C**heese **O**ver **F**ried **G**reen **S**pinach. The first letter of each word can trigger the biological term, as in: Kingdom, Phylum, Class, Order, Family, Genus, Species. This is called a **mnemonic** and is an example of an artificially imposed pattern. Another type of mnemonic is one that helps you to remember a rule. For example, 'tea with two sugars' is a mnemonic for the correct number of t's and s's when spelling 'potassium'.

Memory Models

There are a variety of models used to explain how memory works. The **magical number seven rule, +/- two model**, described opposite, was developed in 1956. In the late 1960s, the **multi-store model** was proposed by Atkinson and Shiffrin. This linked short- and long-term memory with a very quick sensory memory, lasting only 1–2 seconds. This model is useful as it suggests a relationship between memory centres for storage and retrieval, together with an explanation as to how repetition helps people to remember things. It also explains why people forget.

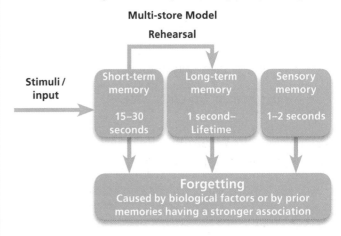

Multi-store Model

Rehearsal

Stimuli / input

Short-term memory	Long-term memory	Sensory memory
15–30 seconds	1 second–Lifetime	1–2 seconds

Forgetting
Caused by biological factors or by prior memories having a stronger association

Biologists are still creating models to explain memory. Models are used to try to explain experimental results. Those that best explain and fit the data are maintained; those that do not are discarded.

However, models will always be limited as they are not the real brain and memory, only a representation of how we think it works.

Exam Practice Questions

B4 **1** On the axes below draw a graph that has the optimum temperature for enzyme activity as being 35°C. **[3]**

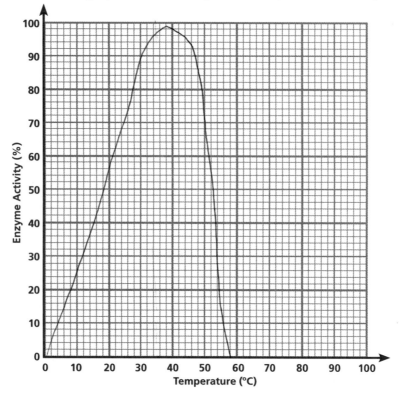

B4 **2** Explain the differences between aerobic and anaerobic respiration. **[6]**

🖉 *The quality of written communication will be assessed in your answer to this question.*

B4 **3** Which of the following are examples of osmosis?
Put ticks (✓) in the boxes next to the **three** correct answers. **[3]**

Water evaporating from leaves ☐

Water moving from plant cell to plant cell ☑

Mixing pure water and sugar solution ☑

Ink spreading through water ☐

A pear losing water in a concentrated salt solution ☐

Water moving from blood to body cells ☐

Sugar absorbed from the intestines into the blood ☐

B6 **4** What is needed before biologists will accept a new theory on memory?
Put ticks (✓) in the boxes next to the **three** correct answers. **[2]**

The theory must…

…have a scientific mechanism. ☐   …be easy to understand. ☐

…be from a qualified biologist. ☐   …be published. ☐

…have data with high variability. ☐   …be repeatable. ☐

B5 5 At what stage do cells in an embryo start to become specialised? [1]

B5 6 This question is about responding to stimuli.
(a) Which way will a seedling in the plant pot grow? [1]

(b) Fill in the words to complete the passage below, which describes what happens when organisms respond to stimuli. [2]

Animals respond to stimuli in order to keep themselves in conditions that will ensure

their .... These responses are coordinated by the ....

B6 7 This question is about reflexes.
(a) Name **three** reflexes shown by newborn babies. [3]

(b) Explain how simple reflexes enable organisms to survive. [6]
✎ *The quality of written communication will be assessed in your answer to this question.*

(c) What is the name given to reflexes that are learned? [1]

HT B4 8 Write the chemical equation for aerobic respiration. [2]

B5 9 Write the complementary mRNA sequence for the following stretch of DNA:

AATAGCCGCCAATTAGGC [1]

B6 10 Four students are talking about Ecstasy.

Ecstasy is a drug that increases the amount of serotonin produced by the brain, which depresses the user.

Ecstasy decreases the amount of serotonin produced in the brain.

Ecstasy blocks serotonin reabsorption sites, making the user feel happy.

Ecstasy is not a drug.

David **Joanne** **Ryan** **Semone**

Who is giving the most accurate statement? [1]

Module C4 (Chemical Patterns)

Theories of atomic structure can be used to explain the properties and behaviour of elements. This module looks at:

- patterns in the properties of the elements
- how to explain the patterns in the properties of the elements
- the development of the periodic table
- the structure of the atom and electron configuration
- ionic bonding
- balanced equations
- how chemists explain the properties of compounds of Group 1 and Group 7 elements.

The Periodic Table

Elements are the 'building blocks' of all materials.

The atoms of each element have a different proton number. The elements are arranged in order of ascending **atomic** (or **proton**) **number**, which gives repeating patterns in the properties of elements.

The Development of the Periodic Table

Knowledge of chemical facts began to grow in the 18th and 19th centuries. As a result, scientists tried to find patterns in order to stop themselves being overwhelmed by the mass of information and to provide a basis for understanding the facts. Many of the early attempts of classification were dismissed by the scientific community as more information emerged.

Three of the most significant developments were made by Döbereiner, Newlands and Mendeleev.

Döbereiner (1829)

Döbereiner arranged elements in groups of three, known as triads. The elements in each triad had similar chemical properties, and the relative atomic mass of the middle one was about halfway between the other two.

For example:
Li – 7; Na – 23; K – 39
Cl – 35.5; Br – 80; I – 127

Atomic mass at that time was measured relative to the mass of a hydrogen atom. It is now measured relative to $\frac{1}{12}$ the mass of a carbon atom.

John Newlands (1864)

Newlands only knew of the existence of 63 elements; many were still undiscovered. He arranged the known elements in order of relative atomic mass and found similar properties amongst every eighth element in the series. This makes sense since the noble gases (Group 0) weren't discovered until 1894.

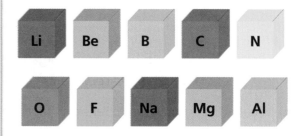

He had noticed some patterns, but the missing elements caused problems. Unfortunately his idea of putting elements into groups of eight had flaws and society greeted his idea with ridicule. The reasons for the flaws were later explained by the discovery of new elements, incorrect data and more complicated patterns after calcium.

Dimitri Mendeleev (1869)

Mendeleev realised that some elements had yet to be discovered, so he left gaps to accommodate their eventual discovery.

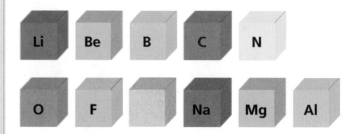

He used his periodic table to predict the existence of other elements. He also challenged some of the previous atomic mass data as being inaccurate.

Today's Periodic Table

Groups

A **vertical column** of elements is called a **group**. For example, lithium (Li), sodium (Na) and potassium (K) are all elements found in Group 1.

Elements in the same group have the same number of electrons in their outer shell (except helium). This number also coincides with the group number. For example, Group 1 elements have one electron in their outer shell, and Group 7 elements have seven electrons in their outer shell. Elements in the same group have similar properties.

Periods

A **horizontal row** of elements is called a **period**. For example, lithium (Li), carbon (C) and neon (Ne) are all elements in the same period.

The period to which an element belongs corresponds to the number of shells of electrons it has. For example, sodium (Na), aluminium (Al) and chlorine (Cl) all have three shells of electrons, so they are found in the third period.

The periodic table can be used as a reference table to obtain important information about the elements. For example:

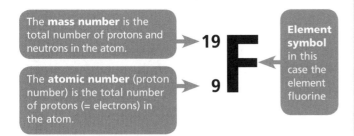

The **mass number** is the total number of protons and neutrons in the atom.

The **atomic number** (proton number) is the total number of protons (= electrons) in the atom.

19 F 9

Element symbol in this case the element fluorine

You can also tell if elements are metals or non-metals by looking at their position in the table. Metals are generally found on the left-hand side of the periodic table and non-metals on the right.

You will be given a copy of the periodic table in the exam. You can find a copy at the back of this book.

Atoms

Chemists use **atoms** to explain the properties of the elements. All substances are made up of atoms (very small particles). Each atom has a small central nucleus, made up of **protons** and **neutrons** (with the exception of hydrogen), which is surrounded by **electrons** arranged in **shells** (or **energy levels**).

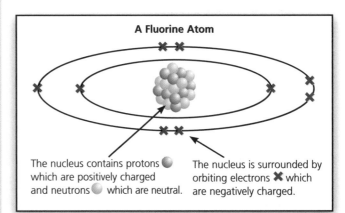

A Fluorine Atom

The nucleus contains protons which are positively charged and neutrons which are neutral.

The nucleus is surrounded by orbiting electrons ✖ which are negatively charged.

Atomic Particle	Relative Mass	Relative Charge
Proton	1	+1
Neutron	1	0
Electron ✖	0 (nearly)	-1

- An atom has the same number of protons as electrons, so the atom as a whole is neutral (i.e. it has no electrical charge).
- A proton has the same mass as a neutron.
- The mass of an electron is negligible, i.e. nearly nothing, when compared to a proton or neutron.
- A substance that contains only one sort of atom is called an element.
- All atoms of the same element have the same number of protons.
- Atoms of different elements have different numbers of protons.
- The elements are arranged in the periodic table in order of increasing atomic (proton) number.

HT You must be able to use the periodic table to work out the number of protons, electrons and neutrons in atoms of the first 20 elements.

Spectroscopy

When some elements are heated, they emit distinctive coloured flames. Lithium, sodium, and potassium compounds can be recognised by the distinctive colours they produce in a flame test.

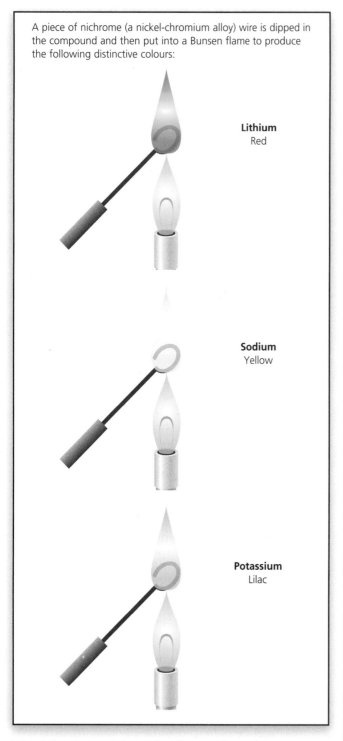

A piece of nichrome (a nickel-chromium alloy) wire is dipped in the compound and then put into a Bunsen flame to produce the following distinctive colours:

Lithium
Red

Sodium
Yellow

Potassium
Lilac

The light emitted from the flame of an element produces a characteristic line spectrum. Each line in the spectrum represents an energy change as excited electrons fall from high energy levels to lower energy levels. The study of spectra has helped chemists to discover new elements. The discovery of some elements depended on the development of new practical techniques like spectroscopy. White light shows a continuous spectrum, but the spectrum of a lithium flame is made up of a series of lines:

Increasing wavelength

Electron Configuration

Electron configuration tells us how the electrons are arranged around the nucleus of an atom in shells (energy levels).

- The electrons in an atom occupy the lowest available shells (i.e. the shells closest to the nucleus).
- The first level (or shell) can only contain a maximum of two electrons.
- The second shell can hold a maximum of eight electrons.
- The third shell can hold up to 18 electrons (however, the last 10 of these electrons are only filled up after the first two in the fourth shell).
- The electron configuration is written as a series of numbers, e.g. oxygen is 2.6; aluminium is 2.8.3; and potassium is 2.8.8.1.

There is a connection between the number of outer electrons and the group. For example, an element with one electron in its outer shell is found in Group 1. You can also deduce the period to which an element belongs from its electron configuration. For example, oxygen is 2.6, which means it is found in the second period; potassium is 2.8.8.1, which means it is found in the fourth period, and so on (i.e. the number of energy levels occupied by electrons equals the number of the period).

HT The chemical properties of an element are determined by its electron arrangement.

Electron Configuration of the First 20 Elements

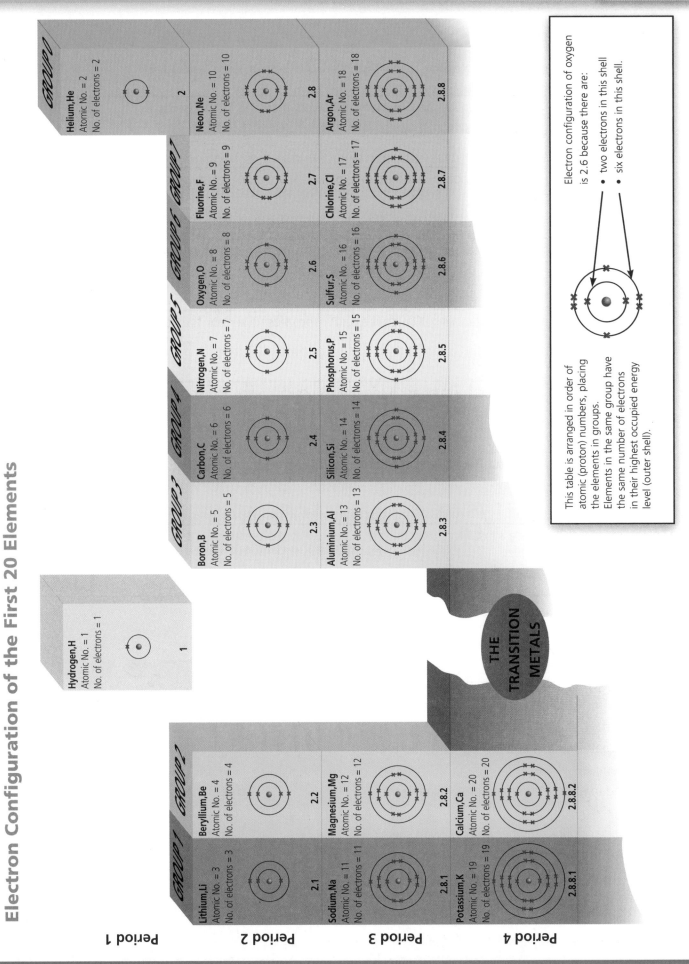

This table is arranged in order of atomic (proton) numbers, placing the elements in groups.
Elements in the same group have the same number of electrons in their highest occupied energy level (outer shell).

Electron configuration of oxygen is 2.6 because there are:
- two electrons in this shell
- six electrons in this shell.

THE TRANSITION METALS

Hydrogen, H Atomic No. = 1 No. of electrons = 1 — 1

GROUP 1
- Lithium, Li — Atomic No. = 3 — No. of electrons = 3 — 2.1
- Sodium, Na — Atomic No. = 11 — No. of electrons = 11 — 2.8.1
- Potassium, K — Atomic No. = 19 — No. of electrons = 19 — 2.8.8.1

GROUP 2
- Beryllium, Be — Atomic No. = 4 — No. of electrons = 4 — 2.2
- Magnesium, Mg — Atomic No. = 12 — No. of electrons = 12 — 2.8.2
- Calcium, Ca — Atomic No. = 20 — No. of electrons = 20 — 2.8.8.2

GROUP 3
- Boron, B — Atomic No. = 5 — No. of electrons = 5 — 2.3
- Aluminium, Al — Atomic No. = 13 — No. of Electrons = 13 — 2.8.3

GROUP 4
- Carbon, C — Atomic No. = 6 — No. of electrons = 6 — 2.4
- Silicon, Si — Atomic No. = 14 — No. of electrons = 14 — 2.8.4

GROUP 5
- Nitrogen, N — Atomic No. = 7 — No. of electrons = 7 — 2.5
- Phosphorus, P — Atomic No. = 15 — No. of electrons = 15 — 2.8.5

GROUP 6
- Oxygen, O — Atomic No. = 8 — No. of electrons = 8 — 2.6
- Sulfur, S — Atomic No. = 16 — No. of electrons = 16 — 2.8.6

GROUP 7
- Fluorine, F — Atomic No. = 9 — No. of electrons = 9 — 2.7
- Chlorine, Cl — Atomic No. = 17 — No. of electrons = 17 — 2.8.7

GROUP 0
- Helium, He — Atomic No. = 2 — No. of electrons = 2 — 2
- Neon, Ne — Atomic No. = 10 — No. of electrons = 10 — 2.8
- Argon, Ar — Atomic No. = 18 — No. of electrons = 18 — 2.8.8

Period 1
Period 2
Period 3
Period 4

Hazard Symbols

Hazardous materials will have one of the following hazard symbols on their packaging.

Toxic	These substances can kill when swallowed, breathed in or absorbed through the skin.	
Oxidising	These substances provide oxygen, which allows other substances to burn more fiercely.	
Harmful	These substances are similar to toxic substances, but they are less dangerous.	
Highly Flammable	These substances will catch fire easily. They pose a serious fire risk.	
Corrosive	These substances attack living tissue, including eyes and skin, and can damage materials.	
Explosive	These substances will explode when they are set alight.	
Environmental Hazard	These substances may present a danger to the environment.	

Safety Precautions

Some common safety precautions are:

- wearing gloves and eye protection, and washing hands after handling chemicals
- using safety screens
- using small amounts and low concentrations of the chemicals
- working in a fume cupboard or ventilating the room
- not eating or drinking when working with chemicals
- not working near naked flames.

Group 1 – The Alkali Metals

There are six metals in Group 1. As we go down the group, the alkali metals become more reactive.

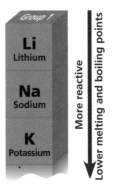

Alkali metals have low melting points. The melting and boiling points decrease as we go down the group.

Physical Properties of the Alkali Metals

Element	Melting Point (°C)	Boiling Point (°C)	Density (g/cm³)
Lithium, Li	180	1340	0.53
Sodium, Na	98	883	0.97
Potassium, K	64	760	0.86
Rubidium, Rb	39	688	1.53
Caesium, Cs	29	671	1.90

Alkali Metal Compounds

When alkali metals react they form compounds that are similar. The reaction becomes more vigorous as we go down the group because the metals are more reactive.

The table below shows the compounds formed when alkali metals react with water, oxygen and chlorine.

Element	+ Water	+ Oxygen	+ Chlorine
Lithium, Li	LiOH	Li_2O	LiCl
Sodium, Na	NaOH	Na_2O	NaCl
Potassium, K	KOH	K_2O	KCl

Reaction of Alkali Metals with Chlorine

Chlorine reacts vigorously with alkali metals to form colourless crystalline salts called metal chlorides, for example:

Lithium + Chlorine ⟶ Lithium chloride

$$2Li_{(s)} + Cl_{2(g)} \longrightarrow 2LiCl_{(s)}$$

You must be able to interpret symbol equations like the one above.

You must be able to write balanced equations for the reactions of the alkali metals and halogens.

Reaction of Alkali Metals with Oxygen

The alkali metals are stored under oil because they react very vigorously with oxygen and water. When freshly cut, they are shiny. However, they quickly tarnish in moist air, go dull and become covered in a layer of metal oxide. For example:

Lithium + Oxygen ⟶ Lithium oxide

$$4Li_{(s)} + O_{2(g)} \longrightarrow 2Li_2O_{(s)}$$

Reaction of Alkali Metals with Water

Lithium, sodium and potassium float on top of cold water (due to their low density). The heat from the reaction is great enough to turn sodium and potassium into liquids. Lithium reacts gently, sodium more aggressively and potassium so aggressively it catches fire.

When alkali metals react with water, a metal hydroxide and hydrogen gas are formed. The metal hydroxide dissolves in water to form an alkaline solution, for example:

Potassium + Water ⟶ Potassium hydroxide + Hydrogen

$$2K_{(s)} + 2H_2O_{(l)} \longrightarrow 2KOH_{(aq)} + H_{2(g)}$$

Follow these steps to check the pH level of the solution formed when an alkali metal is added to water.

1. Put some universal indicator into a beaker containing water (H_2O). The universal indicator will be green, indicating that the water is neutral (pH 7).

2. Put a small piece of potassium into the beaker. It will react with the water and give off hydrogen gas (H_2).

The hydrogen gas ignites and burns

Potassium

3. When it has finished reacting, the beaker will contain potassium hydroxide solution, i.e. KOH(aq). The solution will now be purple, which indicates it is alkaline.

Group 1 – The Alkali Metals (Cont.)

Since all the reactions of alkali metals are similar, a general equation or formula is sometimes used. In each case M refers to the alkali metal.

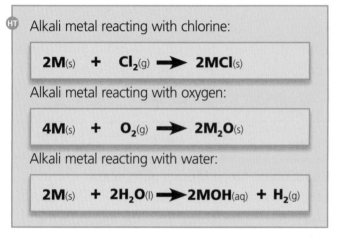

HT

Alkali metal reacting with chlorine:

$$2M_{(s)} + Cl_{2(g)} \rightarrow 2MCl_{(s)}$$

Alkali metal reacting with oxygen:

$$4M_{(s)} + O_{2(g)} \rightarrow 2M_2O_{(s)}$$

Alkali metal reacting with water:

$$2M_{(s)} + 2H_2O_{(l)} \rightarrow 2MOH_{(aq)} + H_{2(g)}$$

The general formula of the chlorides of the alkali metals is MCl, and the general formula of the hydroxides of the alkali metals is MOH. You need to be able to interpret symbols like these, as well as symbol equations.

The following hazard symbols are found on alkali metals and their compounds:

Chemical	Hazard Symbol	
Lithium	🔥	🧪
Lithium chloride	✖	
Sodium	🔥	🧪
Sodium hydroxide	🧪	
Potassium	🔥	🧪

Therefore when working with Group 1 metals, the following precautions should be taken:
- Use small amounts of very dilute concentrations.
- Wear safety glasses and use safety screens.
- Watch teacher demonstrations carefully.
- Avoid working near naked flames.

Group 7 – The Halogens

There are five non-metals in Group 7.

At room temperature and room pressure, chlorine is a green gas, bromine is an orange liquid and iodine is a purple / dark grey solid. When heated, bromine forms a brown gas and iodine a pale purple gas.

Chlorine is used to sterilise water and to make pesticides and plastics.

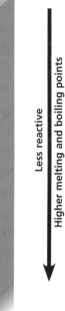

Iodine is used as an antiseptic to sterilise wounds.

All halogens consist of **diatomic molecules** (i.e. they only exist in pairs of atoms), e.g. Cl_2, Br_2 and I_2. They can be used to bleach dyes and kill bacteria in water.

Physical Properties of Halogens

The physical properties of the halogens alter as we go down the group. The table below shows their melting points, boiling points and densities:

Element	Melting Point (°C)	Boiling Point (°C)	Density (g/cm³)
Fluorine, F	-220	-188	0.0016
Chlorine, Cl	-101	-34	0.003
Bromine, Br	-7	59	3.12
Iodine, I	114	184	4.95
Astatine, At	302	337	7 (estimated)

- The melting points and boiling points of the halogens increase as we go down the group.
- The densities increase as we go down the group. (However, due to the unstable nature of astatine, its density is estimated.)

Hazards of the Halogens

The halogens carry the following hazard symbols:

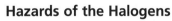

Element	Hazard Symbol
Chlorine, Cl	
Bromine, Br	
Iodine, I	

Therefore when working with halogens, the following precautions should be taken:

- Wear safety glasses.
- Work in a fume cupboard.
- Make sure the room is well ventilated.
- Use small amounts of very dilute concentrations.
- Avoid working near naked flames.
- Watch teacher demonstrations carefully.

Displacement Reactions of Halogens

A more reactive halogen will displace a less reactive halogen from an aqueous solution of its salt. Therefore, chlorine will displace both bromine and iodine, while bromine will displace iodine.

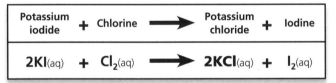

Potassium iodide	+	Chlorine	⟶	Potassium chloride	+	Iodine
$2KI_{(aq)}$	+	$Cl_{2(aq)}$	⟶	$2KCl_{(aq)}$	+	$I_{2(aq)}$

Halogen Compounds

When halogens react, they form compounds that are similar. The reactivity decreases as we go down the group. For example, this is seen clearly when halogens react with the alkali metals or iron. The reaction between chlorine and iron is more vigorous than that between iodine and iron.

The table below shows the compounds that are formed when halogens react with Group 1 metals.

	Chlorine	Bromine	Iodine
Lithium	LiCl	HT LiBr	LiI
Sodium	NaCl	HT NaBr	NaI
Potassium	KCl	HT KBr	KI

HT Trends in Group 1

Alkali metals have similar properties because they have the same number of electrons in their outer shell, i.e. the highest occupied energy level contains one electron.

The alkali metals become more reactive as we go down the group because the outer electron shell is further away from the influence of the nucleus and so an electron is lost more easily.

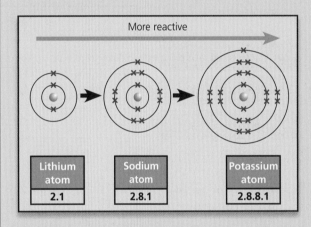

Trends in Group 7

The halogens have similar properties because they have the same number of electrons in their outer shell, i.e. the highest occupied energy level contains seven electrons.

The halogens become less reactive down the group because the outer electron shell is further away from the influence of the nucleus, so an electron is gained less easily.

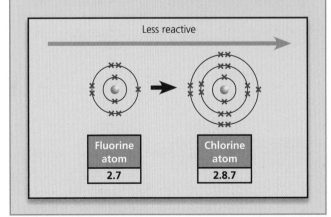

Balanced Equations

The total mass of the products of a chemical reaction is always equal to the total mass of the reactants.

This is because the products of a chemical reaction are made up from the atoms of the reactants – no atoms are lost or made. So, chemical symbol equations must always be balanced: there must be the same number of atoms of each element on the reactant side of the equation as there is on the product side.

Writing Balanced Equations

Follow these steps to write a balanced equation:

1. Write a word equation for the chemical reaction.
2. Substitute in formulae for the elements or compounds involved.
3. Balance the equation by adding numbers in front of the reactants and / or products.
4. Write a balanced symbol equation.

Example 1

Write a word equation

	Reactants			→	Products
	Magnesium	+	Oxygen	→	Magnesium oxide

Substitute in formulae

$Mg(s)$ + $O_2(g)$ → $MgO(s)$

Balance the equation

- There are two **O**s on the reactant side, but only one **O** on the product side. We need to add another **MgO** to the product side to balance the **O**s.
- We now need to add another **Mg** on the reactant side to balance the **Mg**s.
- There are two **Mg** atoms and two **O** atoms on each side – **it is balanced**.

Write a balanced symbol equation

$2Mg(s)$ + $O_2(g)$ → $2MgO(s)$

Example 2

	Reactants			→	Products
Word equation	Sodium	+	Water	→	Sodium hydroxide + Hydrogen
Substitute in formulae	$Na(s)$	+	$H_2O(l)$	→	$NaOH(aq)$ + $H_2(g)$
Balanced symbol equation	$2Na(s)$	+	$2H_2O(l)$	→	$2NaOH(aq)$ + $H_2(g)$

This means that

Two atoms of sodium which are solid	and	Two molecules of water which are liquid	produce	Two sodium hydroxides in aqueous solution	and	One molecule of hydrogen which is a gas

(s), (l), (aq) and (g) are the state symbols

The Properties of Compounds

Chemists use their observations to develop theories to explain the properties of different compounds. For example, experiments show that molten compounds of metals with non-metals, such as lithium chloride, conduct electricity.

It can, therefore, be concluded that there must be charged particles in molten compounds. These particles are known as **ions**.

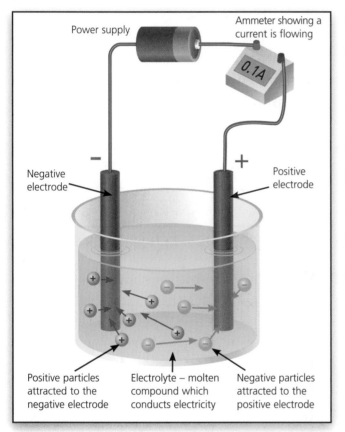

Power supply

Ammeter showing a current is flowing

0.1A

– Negative electrode

+ Positive electrode

Positive particles attracted to the negative electrode

Electrolyte – molten compound which conducts electricity

Negative particles attracted to the positive electrode

Ions

If an atom loses or gains one or more electrons, it will carry an overall charge because the proton and electron numbers are no longer equal. When this happens, the atom becomes an ion.

If the ion has been formed by an atom losing electrons, it will have an overall positive (+) charge because it now has more protons than electrons. It is called a **cation**, e.g. Na^+. If the ion has been formed by an atom gaining electrons, it will have an overall negative (–) charge because it now has more electrons than protons. It is called an **anion**, e.g. Cl^-.

The Ionic Bond

An ionic bond occurs between a metal and a non-metal and involves the transfer of electrons from one atom to another to form electrically charged ions.

Each electrically charged ion has a complete outer energy level or shell, i.e. the first shell has two electrons and each outer shell has eight electrons. Compounds of Group 1 metals and Group 7 elements are **ionic compounds**.

Example 1

Sodium and chlorine bond ionically to form sodium chloride, NaCl. The sodium (Na) atom has one electron in its outer shell which is transferred to the chlorine (Cl) atom, so they both have eight electrons in their outer shell. The atoms become ions, Na^+ and Cl^-, and the compound formed is sodium chloride, NaCl.

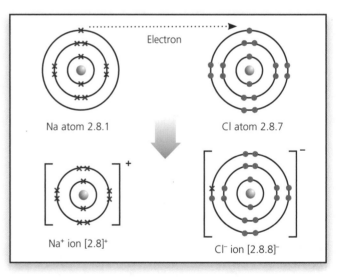

Electron

Na atom 2.8.1

Cl atom 2.8.7

Na^+ ion $[2.8]^+$

Cl^- ion $[2.8.8]^-$

The positive ion and the negative ion are then electrostatically attracted to each other to form a giant crystal lattice.

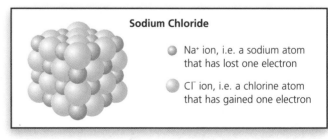

Sodium Chloride

● Na^+ ion, i.e. a sodium atom that has lost one electron

● Cl^- ion, i.e. a chlorine atom that has gained one electron

Sodium chloride has a high melting point and dissolves in water. It conducts electricity when it is in solution or is molten, but not when it is a solid.

Example 2

Sodium and oxygen bond ionically to form sodium oxide, Na_2O. Each sodium (Na) atom has one electron in its outer shell. An oxygen (O) atom wants two electrons, therefore two Na atoms are needed. The atoms become ions (Na^+, Na^+ and O^{2-}) and the compound formed is sodium oxide, Na_2O.

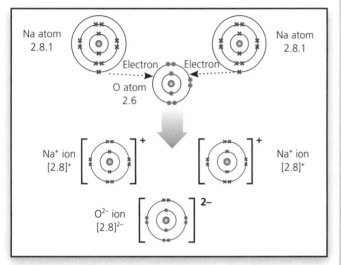

Example 3

Magnesium and oxygen bond ionically to form magnesium oxide, MgO. The magnesium (Mg) atom has two electrons in its outer shell. The oxygen atom only has six electrons in its outer shell, so the two electrons from the Mg atom are transferred to the O atom, so they both have eight electrons in their outer shell. The atoms become ions (Mg^{2+} and O^{2-}) and the compound formed is magnesium oxide, MgO.

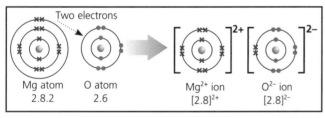

Many ionic compounds have properties similar to those of sodium chloride. They form crystals because the ions are arranged into a regular lattice.

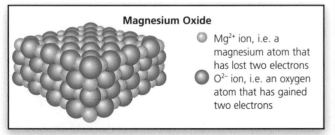

Magnesium Oxide

- Mg^{2+} ion, i.e. a magnesium atom that has lost two electrons
- O^{2-} ion, i.e. an oxygen atom that has gained two electrons

There is a strong force of attraction between the ions, which takes a lot of energy to break, so ionic compounds have high melting and boiling points. When the compound is molten or dissolved, the forces of attraction are weakened so much that the charged ions are free to move. This means they can conduct electricity.

ⓗ Deducing the Formula of an Ionic Compound

If you know the charge on both ions, you can work out the formula of an ionic compound. If you know the formula and the charge on one of the ions, you can work out the charge on the other ion. This is because all ionic compounds are neutral substances where the charge on the positive ion(s) is equal to the charge on the negative ion(s).

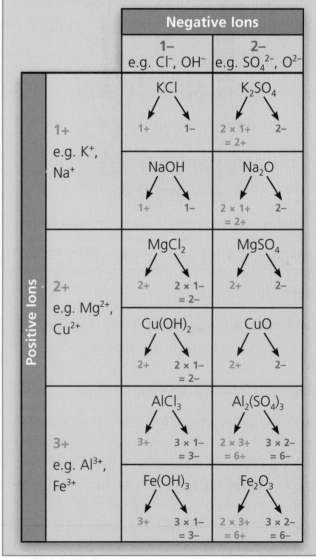

Module C5 (Chemicals of the Natural Environment)

We can get a better understanding of the impact human activity can have on the natural environment by knowing more about the chemicals that make up our planet. This module looks at:

- the structure and properties of the chemicals found in the atmosphere, hydrosphere and lithosphere
- how to extract useful metals from minerals
- the structure and properties of metals.

The Earth's Resources

1 Atmosphere – a layer of gases surrounding the Earth. It is made up of the elements nitrogen (N), oxygen (O), traces of argon (Ar) and some compounds, e.g. carbon dioxide (CO_2) and water vapour (H_2O).

2 Hydrosphere – all the water on the Earth, including oceans, seas, rivers, lakes and underground reserves. The water contains dissolved compounds.

3 Lithosphere – the rigid outer layer of the Earth made up of the crust and the part of the mantle just below it. It is a mixture of minerals, such as silicon dioxide (SiO_2). Abundant elements in the lithosphere include silicon (Si), oxygen (O) and aluminium (Al).

Chemicals of the Atmosphere

Chemical (% composition)	2D Molecular Diagram	3D Molecular Diagram	Boiling Point (°C)	Melting Point (°C)
Oxygen, O_2 (21%)	O=O		-182.9	-218.3
Nitrogen, N_2 (78%)	N≡N		-195.8	-210.1
Carbon dioxide, CO_2 (varies, approx 0.035%)	O=C=O		-78.0	Sublimes (no liquid state)
Water vapour, H_2O (varies from 0–4%)	H–O–H		100.0	0
Argon, Ar (0.93%)	Ar		-185.8	-189.3

The chemicals that make up the atmosphere consist of non-metal elements and molecular compounds made up from non-metal elements.

From the information in the table above we can deduce that the molecules (with the exception of water) that make up the atmosphere are gases at 20°C because they have very low boiling points, i.e. they boil below 20°C. This can be explained by looking at the structure of the molecules. Molecular compounds have strong covalent bonds between the atoms that make up the compound but only weak forces of attraction between the small molecules.

So, for example, gases consist of small molecules with weak forces of attraction between the molecules. Only small amounts of energy are needed to break these forces, which allows the molecules to move freely through the air.

Very weak forces between the molecules

The atoms within molecules (e.g. hydrogen) are connected by strong covalent bonds.

HT In covalent bonds, electrons are shared between the nuclei of two atoms. This causes a strong, electrostatic attraction between the nuclei and shared electrons.

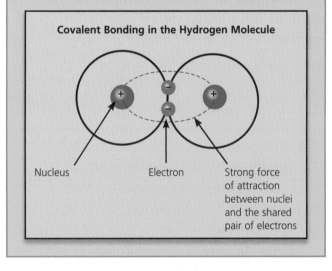

Covalent Bonding in the Hydrogen Molecule

Nucleus   Electron   Strong force of attraction between nuclei and the shared pair of electrons

Unlike ionic compounds, pure molecular compounds do not conduct electricity because their molecules are not charged.

Chemicals of the Hydrosphere

Seawater in the hydrosphere is 'salty' because it contains dissolved ionic compounds. Examples of dissolved ionic compounds are:

- sodium chloride, $NaCl$
- magnesium chloride, $MgCl_2$
- magnesium sulfate, $MgSO_4$
- sodium sulfate, Na_2SO_4
- potassium chloride, KCl
- potassium bromide, KBr.

Examples of solid ionic compounds and how their physical properties relate to their giant structure were given in Module C4.

> **HT** When given a table of charges, you must be able to work out the formulae for the following ionic compounds: sodium chloride, magnesium chloride, sodium sulfate, magnesium sulfate, potassium chloride and potassium bromide.

The Water Molecule

Water has some unexpected properties. For example, the table on the previous page shows that the boiling point of water is 100°C. This is a much higher boiling point than for the other molecules listed. Water is also a good **solvent** for salts. The properties can be explained by its structure.

The water molecule is bent, because the electrons in the covalent bond are nearer to the oxygen atom than the hydrogen atoms. The result is a **polar molecule**.

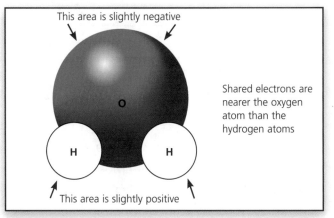

This area is slightly negative

Shared electrons are nearer the oxygen atom than the hydrogen atoms

O

H   H

This area is slightly positive

The small charges on the atoms mean that the forces between the molecules are slightly stronger

than in other covalent molecules. Therefore, more energy is needed to separate them.

The small charges also help water to **dissolve** ionic compounds as the water molecules attract the charges on the ions. The ions can then move freely through the liquid.

Dissolving a Crystal Lattice

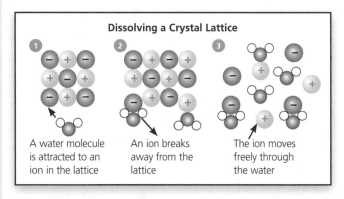

| A water molecule is attracted to an ion in the lattice | An ion breaks away from the lattice | The ion moves freely through the water |

Identification of Ions

Ions in ionic compounds can be detected and identified because they have distinct properties and they form chemicals with distinct properties. For example, an insoluble compound may **precipitate** on mixing two solutions of ionic compounds. This technique is often used to identify metal ions. In the example below, a white precipitate of calcium hydroxide is formed (as well as sodium chloride solution). We can see how the precipitate is formed by considering the ions involved:

$Ca^{2+}_{(aq)}$	+	$2OH^-_{(aq)}$	$\longrightarrow$	$Ca(OH)_{2(s)}$
from the calcium chloride		from the sodium hydroxide		white precipitate

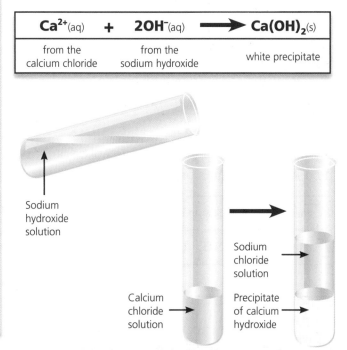

Sodium hydroxide solution

Sodium chloride solution

Calcium chloride solution

Precipitate of calcium hydroxide

Identification of Ions (Cont.)

The table below shows the colour of the precipitate formed when certain metal ions are mixed with alkali.

Metal Ion + Sodium Hydroxide		Precipitate Formed	Colour of Precipitate
Aluminium	Al^{3+}(aq)	Aluminium hydroxide	White
Calcium	Ca^{2+}(aq)	Calcium hydroxide	White
Magnesium	Mg^{2+}(aq)	Magnesium hydroxide	White
Copper	Cu^{2+}(aq)	Copper hydroxide	Blue
Iron(II)	Fe^{2+}(aq)	Iron(II) hydroxide	Green
Iron(III)	Fe^{3+}(aq)	Iron(III) hydroxide	Brown

HT You must be able to write ionic equations for all the precipitate reactions and other reactions given in this module, when provided with appropriate information.

In the oceans, dissolved calcium ions and carbonate ions combine to form a precipitate of calcium carbonate (limestone), which falls to the ocean floor.

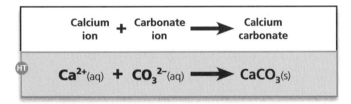

$$Ca^{2+}\text{(aq)} + CO_3{}^{2-}\text{(aq)} \longrightarrow CaCO_3\text{(s)}$$

In order to identify a negative ion, a range of different tests can be carried out. These involve adding a reagent to the unknown sample, which reacts with the ions to form an insoluble salt.

Sulfate Ions

To identify the presence of a sulfate ion, add barium chloride solution and dilute hydrochloric acid to the suspected sulfate solution.

A white precipitate of barium sulfate will be produced if a sulfate is present.

Barium ion + Sulfate ion ⟶ Barium sulfate

HT $Ba^{2+}\text{(aq)} + SO_4{}^{2-}\text{(aq)} \longrightarrow BaSO_4\text{(s)}$

White precipitate

Chloride, Bromide and Iodide Ions

To identify the presence of a chloride, bromide or iodide ion, add silver nitrate solution and nitric acid to the suspected halide solution. A white precipitate will form if silver chloride is present, a cream precipitate for silver bromide, and a yellow precipitate for silver iodide.

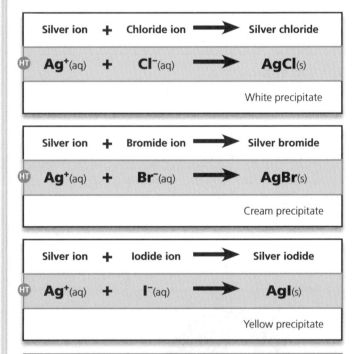

Solubility of Ionic Compounds

In general, most ionic compounds are soluble in water, but as you have seen there are some exceptions. Given information about solubility, you must be able to predict which chemicals are likely to precipitate out when mixing solutions of ionic compounds.

Testing for Carbonates

Testing with Acids

Carbonates react with dilute acids to form carbon dioxide gas (and a salt and water). For example, if we add calcium carbonate to dilute hydrochloric acid then the carbonate will 'fizz' as it reacts with the acid, giving off carbon dioxide.

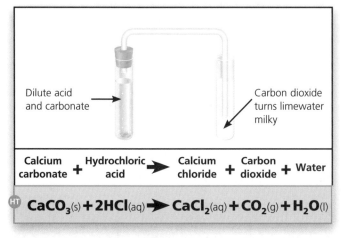

| Dilute acid and carbonate | | | Carbon dioxide turns limewater milky |

Calcium carbonate **+** Hydrochloric acid **→** Calcium chloride **+** Carbon dioxide **+** Water

HT $CaCO_3(s) + 2HCl(aq) \rightarrow CaCl_2(aq) + CO_2(g) + H_2O(l)$

Thermal Decomposition

When copper carbonate and zinc carbonate are heated, a thermal decomposition reaction takes place. This results in a distinctive colour change, which enables the two compounds to be identified.

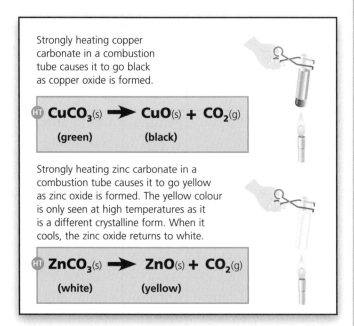

Strongly heating copper carbonate in a combustion tube causes it to go black as copper oxide is formed.

HT $CuCO_3(s) \rightarrow CuO(s) + CO_2(g)$

(green)   (black)

Strongly heating zinc carbonate in a combustion tube causes it to go yellow as zinc oxide is formed. The yellow colour is only seen at high temperatures as it is a different crystalline form. When it cools, the zinc oxide returns to white.

HT $ZnCO_3(s) \rightarrow ZnO(s) + CO_2(g)$

(white)   (yellow)

You must be able to interpret the information given on pages 45–46 to work out which ions are present in an unknown sample when you are given a test with the result.

Chemicals of the Lithosphere

Element	Abundance in Lithosphere (ppm)
Oxygen, O	455 000
Silicon, Si	272 000
Aluminium, Al	83 000
Iron, Fe	62 000
Calcium, Ca	46 600
Magnesium, Mg	27 640
Sodium, Na	22 700
Potassium, K	18 400
Titanium, Ti	6320
Hydrogen, H	1520

N.B. You may be asked to interpret data like this in your exam.

The table shows that the three most abundant elements in the Earth's crust are oxygen, silicon and aluminium. Much of the silicon and oxygen is present as the compound silicon dioxide (SiO_2).

Silicon Dioxide

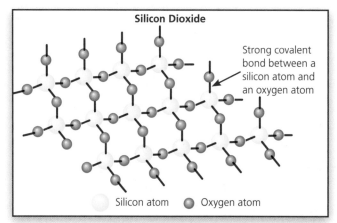

Silicon Dioxide

Strong covalent bond between a silicon atom and an oxygen atom

○ Silicon atom   ● Oxygen atom

Silicon dioxide forms a **giant covalent structure**, where each silicon atom is covalently bonded to four oxygen atoms. Each oxygen atom is bonded to two silicon atoms. The result is a very strong, rigid three-dimensional structure, which is very difficult to break down. Silicon dioxide does not conduct electricity because there are no ions or free electrons in the structure. It does not dissolve in water because there are no charges to attract the water molecules. It also has high melting and boiling points.

Chemicals of the Lithosphere (Cont.)

Silicon dioxide exists in different forms, such as **quartz** in granite, and it is the main constituent of sandstone. Amethyst is a form of quartz that is used as a **gemstone**. It is cut, polished and used in jewellery. The violet colour comes from traces of manganese oxide and iron oxides found in the quartz. Some gemstones are very valuable because of their rarity, hardness and shiny appearance.

Carbon

Carbon is another example of a mineral that forms a giant covalent structure. Two forms of carbon are:
- diamond
- graphite.

Diamond

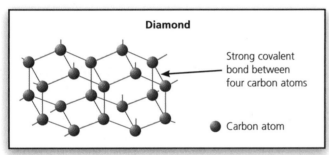

Diamond

Strong covalent bond between four carbon atoms

● Carbon atom

Diamond has a large number of covalent bonds, which means it has very high melting and boiling points. Each carbon atom is covalently bonded to four other carbon atoms, resulting in a very strong, rigid, three-dimensional structure that is difficult to break down. Diamond is insoluble because there are no charges to attract water molecules. It does not conduct electricity because there are no ions or free electrons in the structure.

Graphite

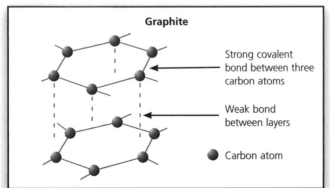

Graphite

Strong covalent bond between three carbon atoms

Weak bond between layers

● Carbon atom

Graphite is a form of carbon that has a giant covalent structure. Each carbon atom is covalently bonded to three other carbon atoms in a layered structure. The layers can slide past each other, making it soft and slippery.

Graphite is also insoluble and has high melting and boiling points, as the carbon atoms in each layer are held together by strong covalent bonds. Graphite can conduct electricity because the electrons forming the weak bonds between the layers are free to move throughout the whole structure.

Using Chemicals of the Lithosphere

By understanding giant covalent structures, we can explain why we use different materials for certain jobs. The table gives some examples:

Element / Compound	Property	Use	Explanation
Carbon – diamond	Very hard	Drill tips	A lot of energy is needed to break the strong covalent bonds between the atoms.
Carbon – graphite	Soft	Pencils	Layers easily removed and stick to paper.
Silicon dioxide	High melting point (1610°C)	Furnace linings	A lot of energy is needed to break the strong covalent bonds between the atoms.
Silica glass	Does not conduct electricity	Insulator in electrical devices	No free electrons or ions to carry electrical charge.

Balancing Equations

All chemical reactions follow the same simple rule: the mass of the reactants is equal to the mass of the products. This means there must be the same number of atoms on both sides of the equation.

You do not have to do anything if the equation is already balanced. If the equation needs balancing, follow this method:

1. Write a number in front of one or more of the formulae. This increases the number of all of the atoms in that formula.
2. Include the state symbols: (s) = solid, (l) = liquid, (g) = gas and (aq) = dissolved in water (aqueous solution).

Example

Balance the reaction between calcium carbonate and hydrochloric acid.

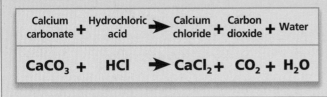

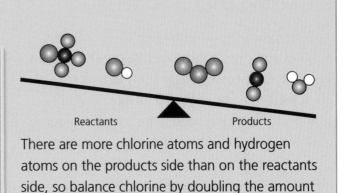

There are more chlorine atoms and hydrogen atoms on the products side than on the reactants side, so balance chlorine by doubling the amount of hydrochloric acid.

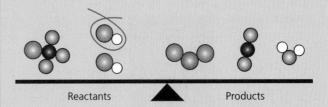

The amount of chlorine and hydrogen on both sides is now equal. This gives you a balanced equation.

$$CaCO_3(s) + 2HCl(aq) \rightarrow CaCl_2(aq) + CO_2(g) + H_2O(l)$$

N.B. When given appropriate information, you must be able to write and balance symbol equations.

Relative Atomic Mass

Atoms are too small for their actual atomic mass to be of much use to us. We therefore use the **relative atomic mass** (**RAM** or **A_r**). This is a number that compares the mass of one atom to the mass of other atoms.

Each element in the periodic table has two numbers. The larger of the two (at the top of the symbol) is the mass number, which also doubles up as the relative atomic mass.

Example

$$^{65}_{30}Zn$$
Zinc

Relative atomic mass of zinc is 65.

Relative Formula Mass

The **relative formula mass** (**RFM** or **M_r**) of a compound is the relative atomic masses of all its elements added together. To calculate RFM we need to know the formula of the compound and the RAM of each of the atoms involved.

Example

Calculate the RFM of water, H_2O.

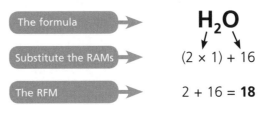

The formula	H_2O
Substitute the RAMs	$(2 \times 1) + 16$
The RFM	$2 + 16 = 18$

The relative formula mass of water is 18.

Extracting Useful Materials

The lithosphere contains many naturally occurring elements and compounds called **minerals**. Ores are rocks that contain varying amounts of minerals, from which metals can be extracted. Sometimes very large amounts of ores need to be mined in order to recover a small percentage of valuable minerals, e.g. copper. The method of extraction depends on the metal's position in the reactivity series.

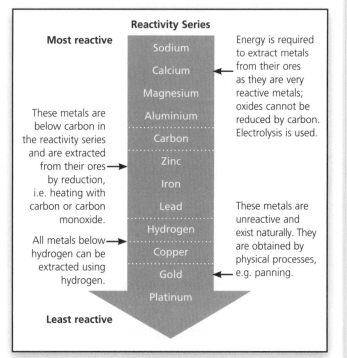

Reactivity Series

Most reactive

Sodium
Calcium
Magnesium
Aluminium
Carbon
Zinc
Iron
Lead
Hydrogen
Copper
Gold
Platinum

Least reactive

Energy is required to extract metals from their ores as they are very reactive metals; oxides cannot be reduced by carbon. Electrolysis is used.

These metals are below carbon in the reactivity series and are extracted from their ores by reduction, i.e. heating with carbon or carbon monoxide.

All metals below hydrogen can be extracted using hydrogen.

These metals are unreactive and exist naturally. They are obtained by physical processes, e.g. panning.

Examples of Extraction by Reduction with Carbon

Zinc can be extracted from zinc oxide by heating it with carbon. Zinc oxide is reduced because it loses oxygen. Carbon is oxidised because it gains oxygen.

Reduction →
Zinc oxide + Carbon ⟶ Zinc + Carbon dioxide
Oxidation →

Iron and copper can also be extracted using carbon reduction:

Reduction →
Iron oxide + Carbon ⟶ Iron + Carbon dioxide
Oxidation →

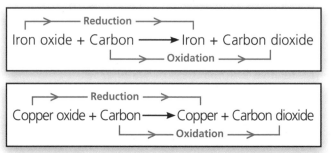

Reduction →
Copper oxide + Carbon ⟶ Copper + Carbon dioxide
Oxidation →

(HT) Calculating a Metal's Mass

If you are given its formula, you can calculate the mass of metal that can be extracted from a substance. Follow these steps:

1. Write down the formula.
2. Work out the relative formula mass.
3. Work out the percentage mass of metal in the formula.
4. Work out the mass of metal.

Example

Find the mass of Zn that can be extracted from 100g of ZnO.

1. The formula is ZnO

2. Relative formula mass = 65 + 16
   = 81

3. Percentage of zinc present

$$= \frac{\text{RAM of Zn}}{\text{RFM of ZnO}} \times 100$$

$$= \frac{65}{81} \times 100 = 80\%$$

4. In 100g of ZnO, there will be $\frac{80}{100} \times 100$
   = **80g of Zn**

If you were given the equation of a reaction, you could find the ratio of the mass of the reactant to the mass of the product.

$$2ZnO + C \longrightarrow 2Zn + CO_2$$

Relative formula mass:

Work out the RFM of each substance

$$(2 \times 81) + 12 = (2 \times 65) + 44$$

$$162 + 12 = 130 + 44$$

$$174 = 174 \checkmark$$

Therefore, 162g of ZnO produces 130g of Zn.

So, 1g of ZnO produces: $\frac{130}{162} = 0.8$g of Zn

and 100g of ZnO produces: $0.8 \times 100 =$ **80g of Zn**.

Metals and the Environment

In order to assess the impact on the environment of extracting and using metals, a life cycle assessment of metal products needs to be carried out. You need to be able to understand and evaluate the sort of data in the table below:

Stage of Life Cycle	Process	Environmental Impact
Making the material from natural raw materials	Mining	• Lots of rock wasted • Leaves a scar on the landscape • Air pollution • Noise pollution
	Processing	• Pollutants caused by transportation • Energy usage
	Extracting the metal	• Electrolysis uses more energy than reduction
Making the product from the material	Manufacturing products	• Energy usage in processing and transportation
Use	Transport	• Pollutants caused by transportation
	Running product	• Energy usage
Disposal	Reuse	• No impact
	Recycle	• Uses a lot less energy than the initial manufacturing
	Throw away	• Landfill sites remove wildlife habitats and are unsightly

Extraction by Electrolysis

Electrolysis is the decomposition of an **electrolyte** (a solution that conducts electricity) using an electric current. The process is used in industry to extract reactive metals from their ores. Ionic compounds will only conduct electricity when their ions are free

to move. This occurs when the compound is either molten or dissolved in solution. During melting of an ionic compound, the electrostatic forces between the charged ions are broken. The crystal lattice is broken down and the ions are free to move.

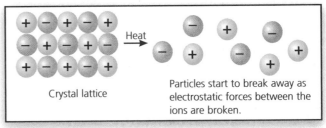

Crystal lattice

Particles start to break away as electrostatic forces between the ions are broken.

Molten lead bromide is a liquid containing positive lead ions and negative bromide ions that are free to move throughout the liquid. When a direct current is passed through the molten salt, the positively charged lead ions are attracted towards the negative electrode. The negatively charged bromide ions are attracted towards the positive electrode. As a result, lead is formed at the negative electrode and bromine at the positive electrode.

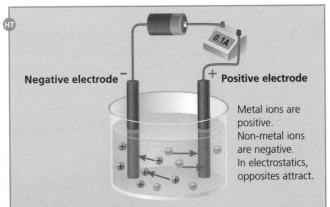

Negative electrode ⁻   ⁺ Positive electrode

Metal ions are positive. Non-metal ions are negative. In electrostatics, opposites attract.

When the ions get to the oppositely charged electrode, they are **discharged** (i.e. they lose their charge).

- The bromide ion loses electrons to the positive electrode to form a bromine atom. The bromine atom then bonds with a second atom to form a bromine molecule.
  $$2Br^- \longrightarrow Br_2 + 2e^-$$
- The lead ions gain electrons from the negative electrode to form a lead atom.
  $$Pb^{2+} + 2e^- \longrightarrow Pb$$

This process completes the circuit as the electrons are exchanged at the electrodes.

Extracting Aluminium by Electrolysis

Aluminium must be obtained from its ore by electrolysis because it is too reactive to be extracted by heating with carbon. (Look at its position in the reactivity series.)

The steps in the process are as follows:

1. Aluminium ore (bauxite) is purified to leave aluminium oxide.
2. Aluminium oxide is mixed with cryolite (a compound of aluminium) to lower its melting point.
3. The mixture of aluminium oxide and cryolite is melted so that the ions can move.
4. When a current passes through the molten mixture, positively charged aluminium ions move towards the negative electrode (the cathode), and aluminium is formed. Negatively charged oxide ions move towards the positive electrode (the anode), and oxygen is formed.
5. This causes the positive electrodes to burn away quickly. They have to be replaced frequently.

The electrolysis of bauxite to obtain aluminium is quite an expensive process due to the cost of the large amounts of electrical energy needed to carry it out. The equation for this reaction is:

Aluminium oxide	$\longrightarrow$	Aluminium	+	Oxygen
$2Al_2O_{3(l)}$	$\longrightarrow$	$4Al_{(l)}$	+	$3O_{2(g)}$

HT The reactions at the electrodes can be written as half equations. This means that we write separate equations for what is happening at each of the electrodes during electrolysis.

At the negative electrode (cathode):

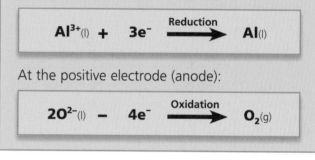

$$Al^{3+}_{(l)} + 3e^- \xrightarrow{\text{Reduction}} Al_{(l)}$$

At the positive electrode (anode):

$$2O^{2-}_{(l)} - 4e^- \xrightarrow{\text{Oxidation}} O_{2(g)}$$

Electrolysis of Aluminium Oxide

Positive carbon electrodes

Carbon lining as negative electrode

Molten aluminium

Oxide ions

Purified aluminium oxide in molten cryolite

Aluminium ions

Steel tank

Tap hole

Properties of Metals

Generally, metals are strong and malleable, have high melting points and can conduct electricity. Their properties determine how each particular metal can be used. For example:

Metal	Properties	Uses
Titanium	• Very strong • Lightweight • Resistant to corrosion	• Replacement hip joints • Bicycles • Submarines
Aluminium	• Malleable • Lightweight • Resistant to corrosion	• Drinks cans • Window frames • Saucepans • Aircraft
Iron	• High melting point • Strong	• Saucepans • Cars
Copper	• Conducts electricity • Conducts heat	• Cables, e.g. kettle cable • Electromagnets • Electrical switches • Saucepans

HT In a metal crystalline structure, the positively charged metal ions are held closely together by a 'sea' of electrons that are free to move.

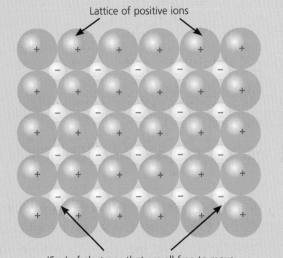

Lattice of positive ions

'Sea' of electrons that are all free to move.

The properties of a metal can be explained by its structure. The force of attraction that keeps the structure together is known as the **metallic bond**. The metallic bond can be used to explain the properties of metals.

Structure	Property and Explanation
The ions are arranged in a lattice form	**Very strong** Metal ions are closely packed in a lattice structure.
	High melting point A lot of energy is needed to break the strong force of attraction between the metal ions and the sea of electrons.
HT Force applied Rows of ions slide over each other. Result: The metal is 'bendy' and can be dented	**Malleable** External forces cause layers of metal ions to move by sliding over other layers.
Moving electrons can carry the electric charge	**Conducts electricity** Electrons are free to move throughout the structure. When an electrical force is applied, the electrons move along the metal in one direction.

Module C6 (Chemical Synthesis)

Chemical synthesis provides the chemicals needed for food processing, health care and many other products. This module looks at:

- chemicals, and why we need them
- acids, alkalis and their reactions
- calculating the mass of products and reactants
- how titrations are used
- measuring the rate of a reaction
- planning, carrying out and controlling a chemical synthesis.

Chemicals

Chemicals are all around us and we depend on them daily. **Chemical synthesis** is the process by which raw materials are made into useful products such as:

- food additives
- fertilisers
- dyestuffs
- pigments
- pharmaceuticals
- cosmetics
- paints.

The chemical industry makes **bulk chemicals**, such as sulfuric acid and ammonia, on a very large scale (millions of tonnes per year). **Fine chemicals**, such as drugs and pesticides, are made on a much smaller scale.

The range of chemicals made in industry and laboratories in the UK is illustrated in the pie chart below.

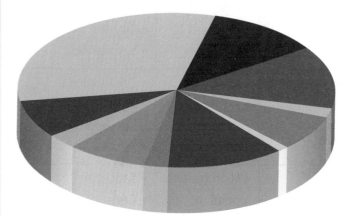

Key:

- Pharmaceuticals 31.5%
- Paint, varnishes and printing inks 8%
- Agrochemicals 3%
- Industrial glass 5%
- Dyes and pigments 3%
- Basic inorganics 2.5%
- Basic organics 12%
- Fertilisers 1%
- Plastic and synthetic rubber 7.5%
- Synthetic fibres 2%
- Other specialities 13%
- Soaps, toiletries and cleaning preparations 11.5%

Many of the raw materials are hazardous, therefore it is important to recognise the standard hazard symbols and understand the necessary precautions that need to be taken (see Module C4).

The pH Scale

The pH scale is a measure of the acidity or alkalinity of an aqueous solution, across a 14-point scale.

Acids are substances that have a pH less than 7.

Bases are the oxides and hydroxides of metals. Those which are soluble are called **alkalis** and they have a pH greater than 7.

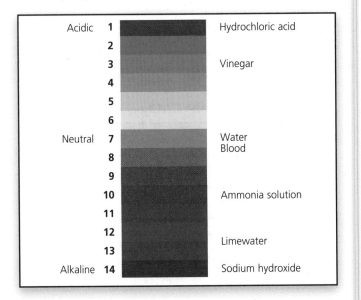

Acidic	**1**	Hydrochloric acid
	2	
	3	Vinegar
	4	
	5	
	6	
Neutral	**7**	Water / Blood
	8	
	9	
	10	Ammonia solution
	11	
	12	
	13	Limewater
Alkaline	**14**	Sodium hydroxide

Indicators such as litmus paper, universal indicator and pH meters are used to detect whether a substance is acidic or alkaline. The pH of a substance is measured using a full range indicator, such as universal indicator solution, or a pH meter.

Universal Indicator Solution

Acidic compounds produce aqueous **hydrogen ions**, $H^+(aq)$, when they dissolve in water.

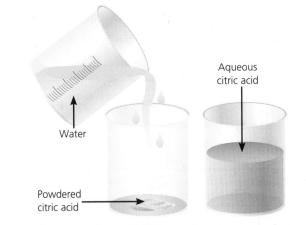

Water

Aqueous citric acid

Powdered citric acid

Common Acids	Formulae to Remember	State at Room Temp.
Citric acid	–	Solid
Tartaric acid	–	Solid
Nitric acid	HNO_3	Liquid
Sulfuric acid	H_2SO_4	Liquid
Ethanoic acid	–	Liquid
Hydrogen chloride (hydrochloric acid)	HCl	Gas

Alkalis produce aqueous **hydroxide ions**, $OH^-(aq)$, when they dissolve in water.

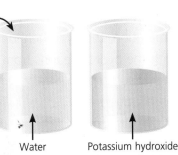

Potassium hydroxide

Water

Potassium hydroxide

Common Alkalis	Formulae to Remember
Sodium hydroxide	NaOH
Potassium hydroxide	–
Magnesium hydroxide	$Mg(OH)_2$
Calcium hydroxide	–

Neutralisation

When an acid and a base are mixed together in the correct amounts, they 'cancel' each other out. This reaction is called **neutralisation** because the solution that remains has a neutral pH of 7.

Acid	+	Base	→	Neutral salt solution	+	Water

Example

Hydrochloric acid	+	Potassium hydroxide	→	Potassium chloride	+	Water

$$HCl_{(aq)} + KOH_{(aq)} \longrightarrow KCl_{(aq)} + H_2O_{(l)}$$

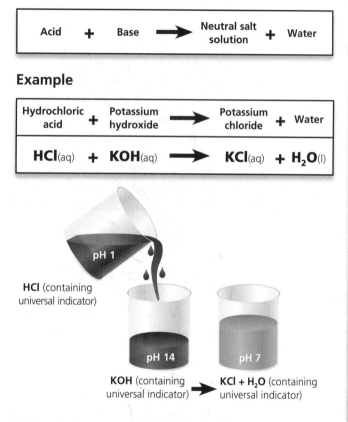

HCl (containing universal indicator) — pH 1

KOH (containing universal indicator) — pH 14

KCl + H₂O (containing universal indicator) — pH 7

N.B. A balanced equation for a chemical reaction shows the relative numbers of atoms and molecules of reactants and products taking part in the reaction.

During neutralisation, the hydrogen ions from the acid react with the hydroxide ions from the alkali to make water. The simplest way of writing a neutralisation equation is:

$$H^+_{(aq)} + OH^-_{(aq)} \longrightarrow H_2O_{(l)}$$

You need to remember this

The salt produced during neutralisation depends on the metal in the base and the acid used. Hydrochloric acid produces chloride salts; sulfuric acid produces sulfate salts; and nitric acid produces nitrate salts.

> **HT** You must be able to write balanced equations to describe the characteristic reactions of acids.

Hydrochloric Acid

Hydrochloric acid	+	Zinc	→	Zinc chloride	+	Hydrogen

$$2HCl_{(aq)} + Zn_{(s)} \rightarrow ZnCl_{2(aq)} + H_{2(g)}$$

Hydrochloric acid	+	Potassium hydroxide	→	Potassium chloride	+	Water

$$HCl_{(aq)} + KOH_{(aq)} \rightarrow KCl_{(aq)} + H_2O_{(l)}$$

Hydrochloric acid	+	Copper oxide	→	Copper chloride	+	Water

$$2HCl_{(aq)} + CuO_{(s)} \rightarrow CuCl_{2(aq)} + H_2O_{(l)}$$

Hydrochloric acid	+	Calcium carbonate	→	Calcium chloride	+	Water	+	Carbon dioxide

$$2HCl_{(aq)} + CaCO_{3(s)} \rightarrow CaCl_{2(aq)} + H_2O_{(l)} + CO_{2(g)}$$

Sulfuric Acid

Sulfuric acid	+	Zinc	→	Zinc sulfate	+	Hydrogen

$$H_2SO_{4(aq)} + Zn_{(s)} \rightarrow ZnSO_{4(aq)} + H_{2(g)}$$

Sulfuric acid	+	Potassium hydroxide	→	Potassium sulfate	+	Water

$$H_2SO_{4(aq)} + 2KOH_{(aq)} \rightarrow K_2SO_{4(aq)} + 2H_2O_{(l)}$$

Sulfuric acid	+	Copper oxide	→	Copper sulfate	+	Water

$$H_2SO_{4(aq)} + CuO_{(s)} \rightarrow CuSO_{4(aq)} + H_2O_{(l)}$$

Sulfuric acid	+	Calcium carbonate	→	Calcium sulfate	+	Water	+	Carbon dioxide

$$H_2SO_{4(aq)} + CaCO_{3(s)} \rightarrow CaSO_{4(aq)} + H_2O_{(l)} + CO_{2(g)}$$

Nitric Acid

Nitric acid	+	Zinc	→	Zinc nitrate	+	Hydrogen

$$2HNO_{3(aq)} + Zn_{(s)} \rightarrow Zn(NO_3)_{2(aq)} + H_{2(g)}$$

Nitric acid	+	Potassium hydroxide	→	Potassium nitrate	+	Water

$$HNO_{3(aq)} + KOH_{(aq)} \rightarrow KNO_{3(aq)} + H_2O_{(l)}$$

Nitric acid	+	Copper oxide	→	Copper nitrate	+	Water

$$2HNO_{3(aq)} + CuO_{(s)} \rightarrow Cu(NO_3)_{2(aq)} + H_2O_{(l)}$$

Nitric acid	+	Calcium carbonate	→	Calcium nitrate	+	Water	+	Carbon dioxide

$$2HNO_{3(aq)} + CaCO_{3(s)} \rightarrow Ca(NO_3)_{2(aq)} + H_2O_{(l)} + CO_{2(g)}$$

Writing Formulae

You need to remember the formulae of the salts listed in this table.

Group of Metal	Salt	Formula to Remember
1	Sodium chloride	NaCl
1	Potassium chloride	KCl
1	Sodium carbonate	Na_2CO_3
1	Sodium nitrate	$NaNO_3$
1	Sodium sulfate	Na_2SO_4
2	Magnesium sulfate	$MgSO_4$
2	Magnesium carbonate	$MgCO_3$
2	Magnesium oxide	MgO
2	Magnesium chloride	$MgCl_2$
2	Calcium carbonate	$CaCO_3$
2	Calcium chloride	$CaCl_2$

HT You should already know how to write formulae for **ionic compounds**. Given the formulae of the salts listed in the table above, you need to be able to work out the charge on each ion in a compound. See Module C4.

It is important to know the formulae of the common gases that occur as covalently bonded (diatomic) molecules.

Gas	Formula to Remember
Chlorine	Cl_2
Hydrogen	H_2
Nitrogen	N_2
Oxygen	O_2

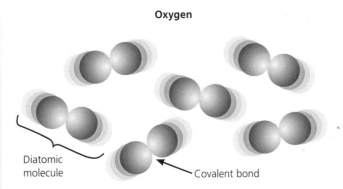

Oxygen

Diatomic molecule

Covalent bond

Energy Changes

When chemical reactions occur, energy is transferred to or from the surroundings. Therefore, many chemical reactions are accompanied by a **temperature change**.

Exothermic Reactions

Exothermic reactions are accompanied by a **temperature rise**. They transfer heat energy to the surroundings, i.e. they give out heat. The combustion of carbon is an example of an exothermic reaction:

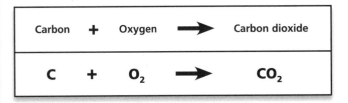

Carbon	+	Oxygen	→	Carbon dioxide
C	+	O_2	→	CO_2

It is not only reactions between fuels and oxygen that are exothermic. Neutralising alkalis with acids and many oxidation reactions also give out heat.

The energy change in an exothermic reaction can be shown using an energy-level diagram. Energy is lost to the surroundings during the reaction, so the products have less energy than the reactants.

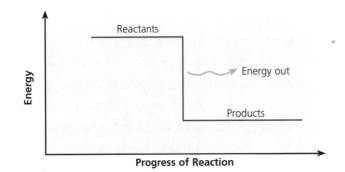

Endothermic Reactions

Endothermic reactions are accompanied by a **fall in temperature**. Heat is transferred from the surroundings, i.e. they take in heat. The reaction between citric acid and sodium hydrogencarbonate is an example of an endothermic chemical reaction. Dissolving ammonium nitrate crystals in water is an example of a physical change that is endothermic.

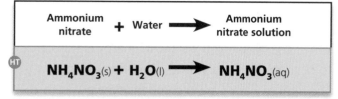

$$NH_4NO_{3(s)} + H_2O_{(l)} \longrightarrow NH_4NO_{3(aq)}$$

Thermal decomposition is also an example of an endothermic reaction.

The energy change in an endothermic reaction can be shown using an energy-level diagram. Energy is taken in during the reaction, so the products have more energy than the reactants.

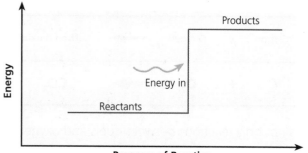

Since energy changes play a major role in chemical reactions, it is very important that these changes are managed during chemical synthesis. For example, there could be a nasty accident if too much energy was given out during a reaction.

Chemical Synthesis

Whenever chemical synthesis takes place, the starting materials (reactants) react to produce new substances (products). The greater the amount of reactants used, the greater the amount of product formed. The **percentage yield** can be calculated by comparing the actual amount of product made (actual yield) with the amount of product you would expect to get if the reaction goes to completion (theoretical yield).

$$\text{Percentage yield} = \frac{\text{Actual yield}}{\text{Theoretical yield}} \times 100$$

There are a number of different stages to any chemical synthesis of an inorganic compound. Look at the flow chart below.

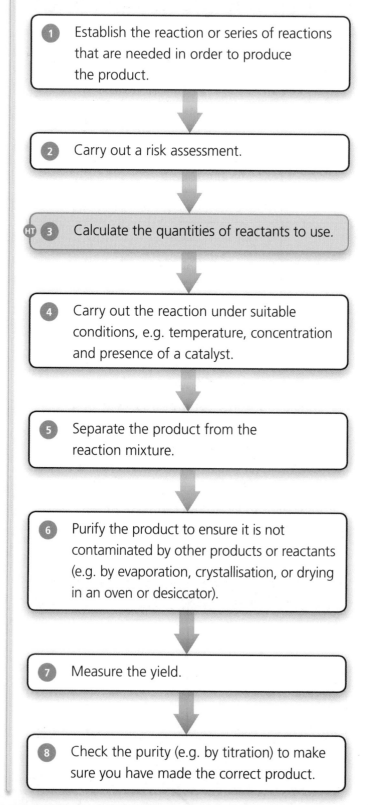

1. Establish the reaction or series of reactions that are needed in order to produce the product.

2. Carry out a risk assessment.

3. Calculate the quantities of reactants to use.

4. Carry out the reaction under suitable conditions, e.g. temperature, concentration and presence of a catalyst.

5. Separate the product from the reaction mixture.

6. Purify the product to ensure it is not contaminated by other products or reactants (e.g. by evaporation, crystallisation, or drying in an oven or desiccator).

7. Measure the yield.

8. Check the purity (e.g. by titration) to make sure you have made the correct product.

Example

The following steps show how to make **sodium chloride**.

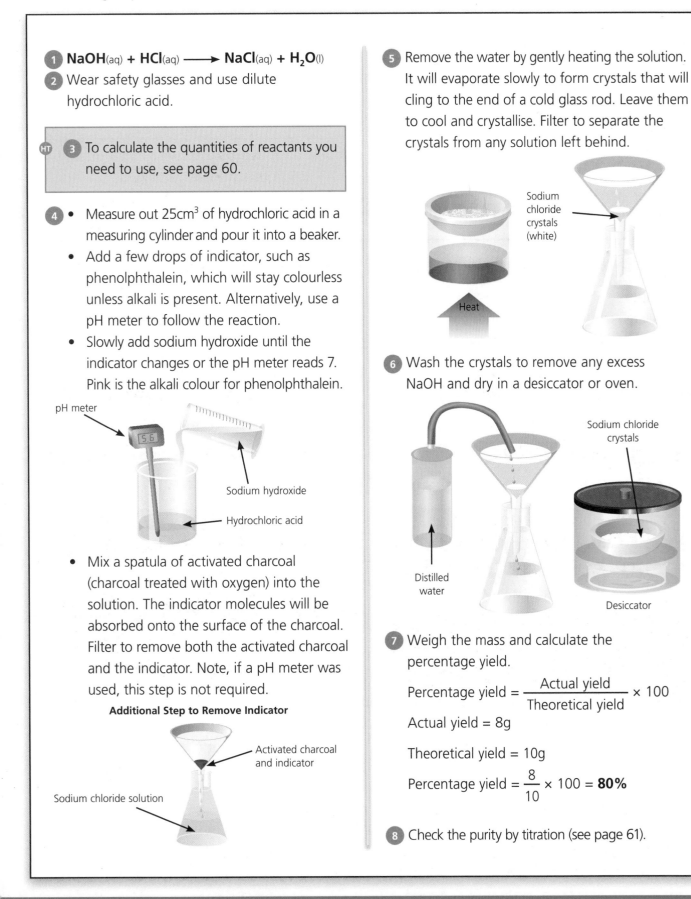

1 $NaOH_{(aq)} + HCl_{(aq)} \longrightarrow NaCl_{(aq)} + H_2O_{(l)}$

2 Wear safety glasses and use dilute hydrochloric acid.

HT 3 To calculate the quantities of reactants you need to use, see page 60.

4
- Measure out 25cm³ of hydrochloric acid in a measuring cylinder and pour it into a beaker.
- Add a few drops of indicator, such as phenolphthalein, which will stay colourless unless alkali is present. Alternatively, use a pH meter to follow the reaction.
- Slowly add sodium hydroxide until the indicator changes or the pH meter reads 7. Pink is the alkali colour for phenolphthalein.

pH meter

5.6

Sodium hydroxide

Hydrochloric acid

- Mix a spatula of activated charcoal (charcoal treated with oxygen) into the solution. The indicator molecules will be absorbed onto the surface of the charcoal. Filter to remove both the activated charcoal and the indicator. Note, if a pH meter was used, this step is not required.

Additional Step to Remove Indicator

Activated charcoal and indicator

Sodium chloride solution

5 Remove the water by gently heating the solution. It will evaporate slowly to form crystals that will cling to the end of a cold glass rod. Leave them to cool and crystallise. Filter to separate the crystals from any solution left behind.

Sodium chloride crystals (white)

Heat

6 Wash the crystals to remove any excess NaOH and dry in a desiccator or oven.

Sodium chloride crystals

Distilled water

Desiccator

7 Weigh the mass and calculate the percentage yield.

$$\text{Percentage yield} = \frac{\text{Actual yield}}{\text{Theoretical yield}} \times 100$$

Actual yield = 8g

Theoretical yield = 10g

$$\text{Percentage yield} = \frac{8}{10} \times 100 = \textbf{80\%}$$

8 Check the purity by titration (see page 61).

Quantity of Reactants and Products

In order to work out how much of each reactant is required to make a known amount of product, you must understand:

- that a balanced equation shows the relative number of atoms or molecules of reactants and products taking part in the reaction
- that the relative atomic mass of an element shows the mass of its atoms relative to the mass of other atoms

- that relative atomic masses of elements can be found from the periodic table
- how to calculate the relative atomic mass (see page 49)
- how to substitute the relative formula masses and data into a given mathematical formula to calculate the reacting masses and/or products from a chemical reaction

HT
- how to work out the ratio of the mass of reactants to the mass of products
- how to apply the ratio to the question.

Calculating a Product's Mass

Example

Calculate how much calcium oxide can be produced from 50kg of calcium carbonate. (Relative atomic masses: Ca = 40, C = 12, O = 16).

> Write down the equation.

$$CaCO_3 \rightarrow CaO + CO_2$$

> Work out the RFM of each substance.

$$40 + 12 + (3 \times 16) \rightarrow (40 + 16) + [12 + (2 \times 16)]$$

> Check that the total mass of the reactants equals the total mass of the products. If they are not the same, check your work.

$$100 \rightarrow 56 + 44 ✓$$

> Since the question only mentions calcium oxide and calcium carbonate, you can now ignore the carbon dioxide. You just need the ratio of mass of reactant to mass of product.

$$100 : 56$$

We need to know how much CaO is produced from 50kg of $CaCO_3$.
100kg of $CaCO_3$ produces 56kg of CaO.
So, 1kg of $CaCO_3$ produces $\frac{56}{100}$ kg of CaO,
and 50kg of $CaCO_3$ produces $\frac{56}{100} \times 50$
= **28kg** of CaO

Calculating a Reactant's Mass

Example

Calculate how much aluminium oxide is needed to produce 540 tonnes of aluminium. (Relative atomic masses: Al = 27, O = 16).

> Write down the equation.

$$2Al_2O_3 \rightarrow 4Al + 3O_2$$

> Work out the RFM of each substance.

$$2[(2 \times 27) + (3 \times 16)] \rightarrow (4 \times 27) + [3 \times (2 \times 16)]$$

> Check that the total mass of the reactants equals the total mass of the products.

$$204 \rightarrow 108 + 96 ✓$$

> Since the question only mentions aluminium oxide and aluminium, you can now ignore the oxygen. You just need the ratio of mass of reactant to mass of product.

$$204 : 108$$

We need to know how much Al_2O_3 is needed to produce 540 tonnes of Al.
204 tonnes of Al_2O_3 produces 108 tonnes of Al.
So, $\frac{204}{108}$ tonnes is needed to produce 1 tonne of Al, and $\frac{204}{108} \times 540$ tonnes is needed to produce 540 tonnes of Al
= **1020 tonnes** of Al_2O_3

Titration

Titration can be used to calculate the concentration of an acid, such as citric acid, by finding out how much alkali is needed to neutralise it.

1 Fill a burette with the alkali sodium hydroxide (the concentration of the alkali must be known) and take an initial reading of the volume.

2 Accurately weigh out a 4g sample of solid citric acid and dissolve it in 100cm³ of distilled water.

3 Use a pipette to measure 25cm³ of the aqueous citric acid and put it into a conical flask.

Add a few drops of the indicator, phenolphthalein, to the conical flask. The indicator will stay colourless.

Place the flask on a white tile under the burette.

4 Add the alkali from the burette to the acid in the flask drop by drop.

- Swirl the flask to ensure it mixes well. Near the end of the reaction, the indicator will start to turn pink.
- Keep swirling and adding the alkali until the indicator is completely pink, showing that the citric acid has been neutralised.
- Record the final burette reading. Work out the volume of alkali added from:
  Volume = final reading – initial reading

Repeat the whole procedure until you get two results that are within +/– 0.05cm³.

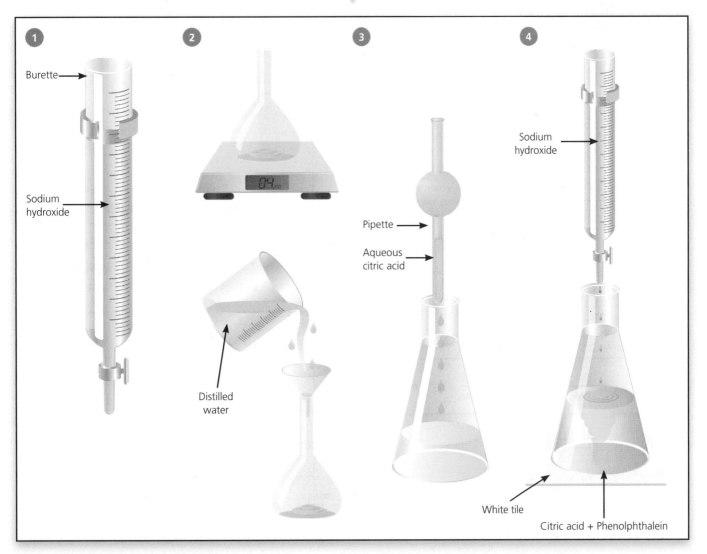

1

Burette

Sodium hydroxide

2

3

Pipette

Aqueous citric acid

4

Sodium hydroxide

Distilled water

White tile

Citric acid + Phenolphthalein

Interpreting Results

Example
Calculate the purity of citric acid used when:

- concentration of sodium hydroxide (NaOH) = 40.0g/dm³
- volume of sodium hydroxide = 8.0cm³
- mass of citric acid = 4.0g
- volume of citric acid solution = 25.0cm³

First, using the formula below, calculate the concentration of citric acid by substituting the values.

$$\text{Concentration of acid} = 3 \times \frac{\text{Volume of conc. NaOH}}{\text{Volume of citric acid}}$$

One unit of citric acid reacts with three units of sodium hydroxide

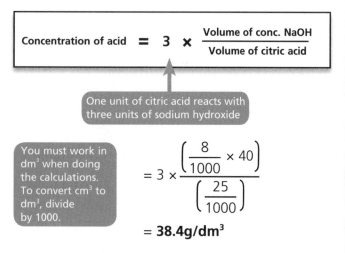

You must work in dm³ when doing the calculations. To convert cm³ to dm³, divide by 1000.

$$= 3 \times \frac{\left(\frac{8}{1000} \times 40\right)}{\left(\frac{25}{1000}\right)}$$

$$= \textbf{38.4g/dm}^3$$

Then, work out the actual mass of citric acid in the sample:

$$\text{Mass} = \text{Concentration} \times \text{Volume}$$

$$= 38.4\text{g/dm}^3 \times \left(\frac{25\text{cm}^3}{1000}\right)$$

$$= \textbf{0.96g}$$

If the mass of citric acid dissolved in 25cm³ is 0.96g, the mass of citric acid dissolved in 100cm³ of water will be 4 × 0.96 = 3.84g

$$\% \text{ of Purity} = \frac{\text{Calculated mass}}{\text{Mass weighed out at start}} \times 100$$

$$= \frac{3.84}{4.0} \times 100$$

$$= \textbf{96\%}$$

N.B. You must be able to substitute the results into a formula to interpret them.

Rates of Reactions

The rate of a **chemical reaction** is the amount of reaction that takes place in a given unit of time. Chemical reactions only occur when the reacting particles collide with each other with sufficient energy to react.

Chemical reactions can proceed at different speeds, e.g. rusting is a slow reaction, whereas burning is a fast reaction.

Measuring the Rate of Reaction

The rate of a chemical reaction can be found in different ways:

1 Weighing the reaction mixture.
If one of the products is a gas, you could weigh the reaction mixture at timed intervals. The mass of the mixture will decrease.

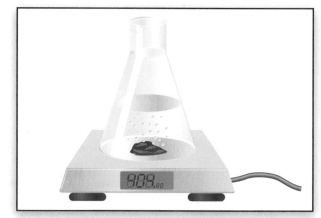

2 Measuring the volume of gas produced.
You could use a gas syringe to measure the total volume of gas produced at timed intervals.

③ Observing the formation of a precipitate.

This can be done by either watching a cross (on a tile underneath the conical flask to see when the cross disappears) in order to measure the formation of a precipitate, or by monitoring a colour change using a light sensor. The light sensor will lead to more reliable and accurate results as there is a definite end point. There are also more data points collected, especially if it is interfaced with a computer.

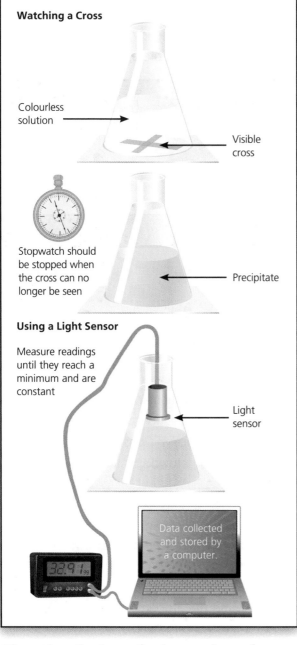

Watching a Cross

Colourless solution

Visible cross

Stopwatch should be stopped when the cross can no longer be seen

Precipitate

Using a Light Sensor

Measure readings until they reach a minimum and are constant

Light sensor

Data collected and stored by a computer.

④ Observing the loss of colour or loss of a precipitate.

This is essentially the opposite of ③.

Analysing the Rate of Reaction

Graphs can be plotted to show the progress of a chemical reaction – there are three things to remember:

① The steeper the line, the faster the reaction.

② When one of the reactants is used up, the reaction stops (line becomes flat).

③ The same amount of product is formed from the same amount of reactants, irrespective of rate.

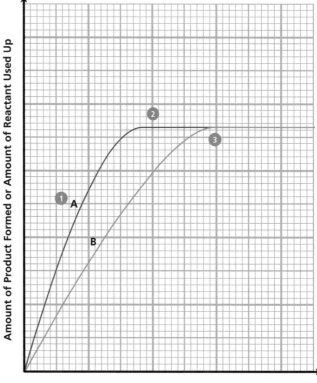

Time (x-axis)

Amount of Product Formed or Amount of Reactant Used Up (y-axis)

Reaction A is faster than reaction B. This could be because:

- the surface area of the solid reactants in reaction A is greater than in reaction B (i.e. smaller particles are used in A)
- the temperature of reaction A is greater than reaction B
- the concentration of the solution in reaction A is greater than in reaction B
- a catalyst is used in reaction A but not in reaction B.

Changing the Rate of Reaction

There are four important factors that affect the rate of reaction – temperature, concentration of dissolved reactants, surface area and the use of a catalyst.

1 Temperature of the Reactants

In a cold reaction mixture, the particles move quite slowly. They will collide less often, with less energy, so fewer collisions will be successful.

In a hot reaction mixture, the particles move more quickly. They will collide more often, with greater energy, so many more collisions will be successful.

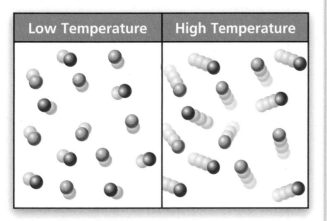

2 Concentration of the Dissolved Reactants

In a low concentration reaction, the particles are spread out. The particles will collide with each other less often, resulting in fewer successful collisions.

In a high concentration reaction, the particles are crowded close together. The particles will collide with each other more often, resulting in many more successful collisions.

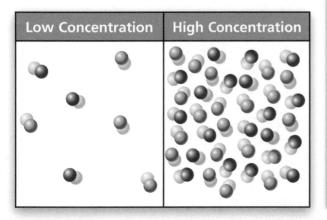

3 Surface Area of Solid Reactants

Large particles (e.g. solid lumps) have a small surface area in relation to their volume, meaning fewer particles are exposed and available for collisions. This means that particles will collide with each other less often, resulting in fewer successful collisions and therefore a slow rate of reaction.

Small particles (e.g. powdered solids) have a large surface area in relation to their volume, so more particles are exposed and available for collisions. This means that particles will collide with each other more often, resulting in many more successful collisions and therefore a faster rate of reaction.

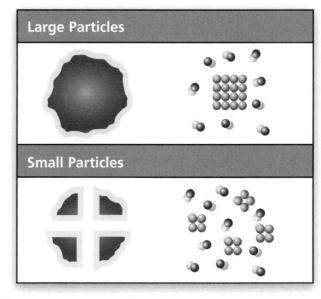

4 Using a Catalyst

A **catalyst** is a substance that increases the rate of a chemical reaction, without being changed during the process. Consider the decomposition of hydrogen peroxide:

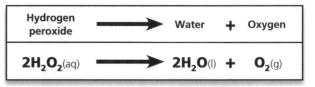

Hydrogen peroxide	$\longrightarrow$	Water	+	Oxygen
$2H_2O_{2(aq)}$	$\longrightarrow$	$2H_2O_{(l)}$	+	$O_{2(g)}$

We can measure the rate of this reaction by measuring the amount of oxygen given off at one-minute intervals. This reaction happens very slowly unless we add a catalyst of manganese(IV) oxide. With a catalyst, plenty of fizzing can be seen as the oxygen is given off.

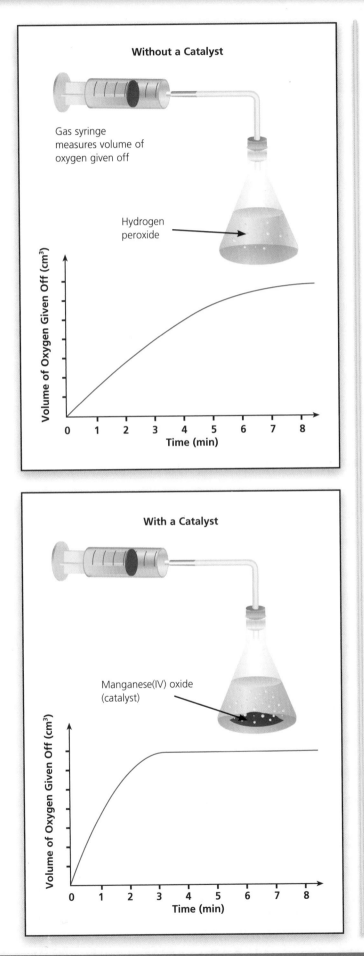

Without a Catalyst

Gas syringe measures volume of oxygen given off

Hydrogen peroxide

Volume of Oxygen Given Off (cm³)

0 1 2 3 4 5 6 7 8
Time (min)

With a Catalyst

Manganese(IV) oxide (catalyst)

Volume of Oxygen Given Off (cm³)

0 1 2 3 4 5 6 7 8
Time (min)

The same amount of gas is given off, but it takes a far shorter time when a catalyst is present.

Catalysts work by lowering the amount of energy needed for a successful collision. They are specific to a particular reaction and are not used up during the reaction. Consequently, only small amounts are needed.

Collision Theory

Chemical reactions only occur when particles collide with each other with sufficient energy. Increasing temperature causes an increase in the kinetic energy of the particles, i.e. they move a lot faster. This results in more energetic collisions happening more frequently. The minimum energy required for a reaction will, therefore, be achieved more often, resulting in a greater rate of reaction.

An increase in concentration or surface area results in more frequent collisions and, therefore, more collisions that are sufficiently energetic for a reaction to occur.

Controlling a Chemical Reaction

When carrying out a chemical synthesis on an industrial scale, there are also economic, safety and environmental factors to consider:

- The rate of manufacture must be high enough to produce a sufficient daily yield of product.
- Percentage yield must be high enough to produce a sufficient daily yield of product.
- A low percentage yield is acceptable, providing the reaction can be repeated many times with recycled starting materials.
- Optimum conditions should be used that give the lowest cost rather than the fastest reaction or highest percentage yield.
- Care must be taken when using any reactants or products that could harm the environment if there was a leak.
- Care must be taken to avoid putting any harmful by-products into the environment.
- A complete risk assessment must be carried out, and the necessary precautions taken.

Exam Practice Questions

C5 **1** These diagrams show the atoms in carbon dioxide gas, nitrogen gas and hydrogen gas.

(a) Write down the formulae of the following compounds. **[2]**

(i) **(ii)**

 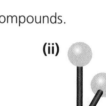

(b) The relative atomic mass of carbon is 12 and the relative atomic mass of oxygen is 16. What is the relative formula mass of carbon dioxide? **[2]**

C5 **2** The lithosphere contains many useful naturally occurring minerals. The process used to extract a metal from its ore depends upon the reactivity of the ore.

(a) Which process best describes the process used to extract iron from its ore? Put a ring around the correct answer. **[1]**

Panning **Reduction** **Electrolysis** **Oxidation**

(b) Before a metal product is manufactured, an analysis is carried out to assess the impact on the environment of extracting and using the metal.

(i) Name three different methods of disposal that might be explored at the final stage of the analysis. **[3]**

(ii) Compare and evaluate the environmental impact of the different methods of disposal identified in part (i). **[4]**

C4 **3** Class Y has been asked to carry out some tests on two different solutions: A and B.

Use the data sheet on page 107 to help you to answer these questions.

(a) Solution A is copper chloride.

(i) Describe what you will see when sodium hydroxide is added to the solution. **[1]**

(ii) Describe what you will see when acidified silver nitrate is added to the solution. **[1]**

(b) Class Y then tests Solution B. Here are the results of the test:

Test	Result
Add dilute sodium hydroxide	Red-brown precipitate forms
Add dilute nitric acid, then silver nitrate	Yellow precipitate forms

What is Solution B? Explain how you came to your conclusion. **[3]**

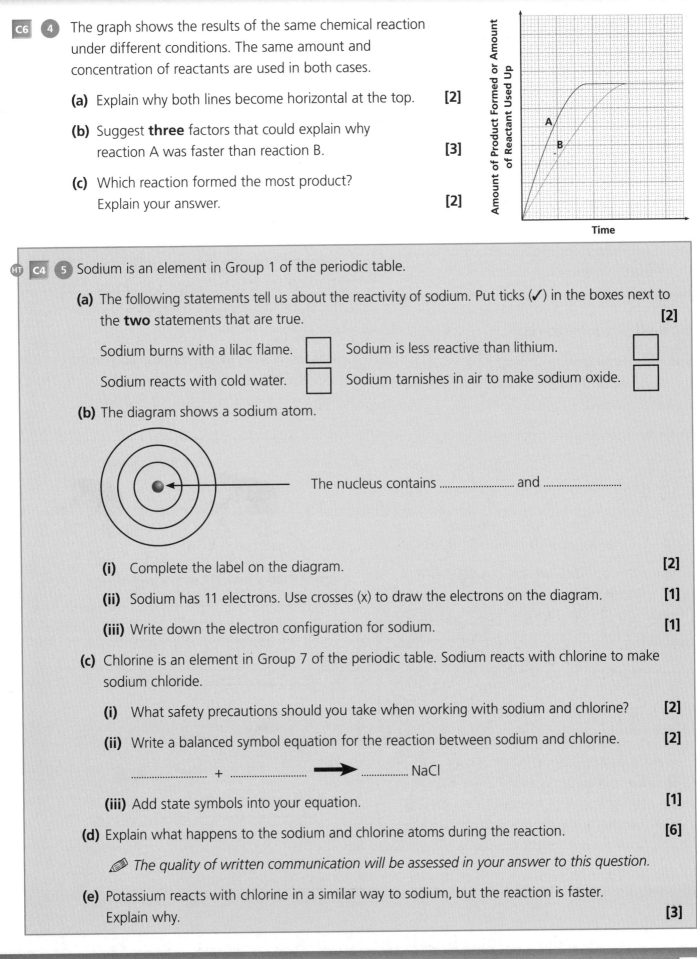

C6 **4** The graph shows the results of the same chemical reaction under different conditions. The same amount and concentration of reactants are used in both cases.

(a) Explain why both lines become horizontal at the top. **[2]**

(b) Suggest **three** factors that could explain why reaction A was faster than reaction B. **[3]**

(c) Which reaction formed the most product? Explain your answer. **[2]**

HT **C4** **5** Sodium is an element in Group 1 of the periodic table.

(a) The following statements tell us about the reactivity of sodium. Put ticks (✓) in the boxes next to the **two** statements that are true. **[2]**

Sodium burns with a lilac flame. ☐ Sodium is less reactive than lithium. ☐

Sodium reacts with cold water. ☐ Sodium tarnishes in air to make sodium oxide. ☐

(b) The diagram shows a sodium atom.

The nucleus contains .... and ....

(i) Complete the label on the diagram. **[2]**

(ii) Sodium has 11 electrons. Use crosses (x) to draw the electrons on the diagram. **[1]**

(iii) Write down the electron configuration for sodium. **[1]**

(c) Chlorine is an element in Group 7 of the periodic table. Sodium reacts with chlorine to make sodium chloride.

(i) What safety precautions should you take when working with sodium and chlorine? **[2]**

(ii) Write a balanced symbol equation for the reaction between sodium and chlorine. **[2]**

.... + ... ⟶ .... NaCl

(iii) Add state symbols into your equation. **[1]**

(d) Explain what happens to the sodium and chlorine atoms during the reaction. **[6]**

🖉 *The quality of written communication will be assessed in your answer to this question.*

(e) Potassium reacts with chlorine in a similar way to sodium, but the reaction is faster. Explain why. **[3]**

An understanding of the rules that govern forces and motion has led to the development of numerous modern technologies, from aeroplanes and spacecraft to sports equipment and theme park rides. This module looks at:

- how we can describe motion
- what forces are
- what the connection is between forces and motion
- how we can describe motion in terms of energy changes.

Speed and Velocity

Speed tells you how far an object will travel in a certain time. However, it does not tell you the direction of travel.

Velocity tells you an object's speed *and* gives an indication of its direction of travel.

HT Example

A lorry is travelling along a straight road. The speed of the lorry is 15m/s (metres per second).

In the morning, travelling in one direction (from the warehouse to the supermarket), the velocity is +15m/s. At night, travelling in the opposite direction (from the supermarket back to the warehouse), the velocity is -15m/s.

It does not matter which direction is called positive or negative as long as opposite directions have opposite signs.

This idea is also used when describing distance; changes in distance in one direction are described as positive, and in the opposite direction they are described as negative.

Calculating Speed

To calculate the speed of an object we need to know two things about its motion:

- The distance it has travelled.
- The time it took to travel that distance.

Speed is calculated using the following formula:

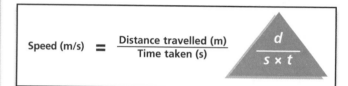

$$\text{Speed (m/s)} = \frac{\text{Distance travelled (m)}}{\text{Time taken (s)}} \qquad \frac{d}{s \times t}$$

The unit for speed is metres per second (m/s), so the distance should be in metres and the time in seconds.

A speed of 4m/s means that the object travels 4 metres every second. At this speed, in 2 seconds, the object would travel a total of 8 metres.

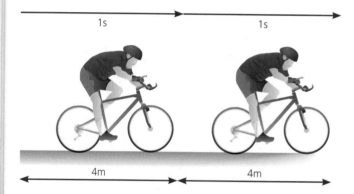

Speed can also be calculated in kilometres per hour, km/h.

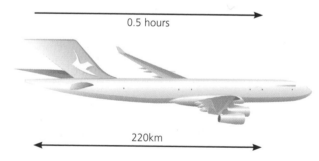

Speed = 440km/h

The formula calculates an average speed over the total distance travelled, even if the speed of an object is not constant.

Example

An object travels 10 metres in 5 seconds. Use the formula to calculate its average speed.

$$\text{Speed} = \frac{\text{Distance travelled}}{\text{Time taken}} = \frac{10m}{5s} = \textbf{2m/s}$$

However, during those 5 seconds the object could have been moving at different speeds.

For example, the object could have moved at 5m/s for 1 second, stopped for 3 seconds, and then moved at 5m/s for 1 second. The object would still have an average speed of 2m/s, because it took 5 seconds to travel 10 metres.

The speed of an object at a particular point in time is called the **instantaneous speed**.

Distance–Time Graphs

A **distance–time graph** shows how the distance travelled by an object changes with time.

The slope or gradient of a distance–time graph is a measure of the speed of the object. The steeper the slope, the greater the speed.

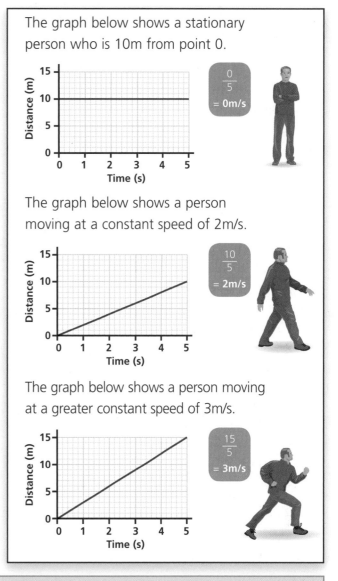

The graph below shows a stationary person who is 10m from point 0.

$$\frac{0}{5} = 0m/s$$

The graph below shows a person moving at a constant speed of 2m/s.

$$\frac{10}{5} = 2m/s$$

The graph below shows a person moving at a greater constant speed of 3m/s.

$$\frac{15}{5} = 3m/s$$

(HT) Distance–time graphs can also be drawn as **displacement–time graphs**, where the displacement of an object is its net distance from its starting point together with an indication of direction.

Example

The displacement–time graph opposite shows how the displacement of a football kicked against a wall changes with time. The ball is kicked, it hits the wall, and then rolls back between the person's legs before coming to a stop.

What is the distance covered by the ball and the displacement of the ball at the following times: 5s, 11s, 22s?

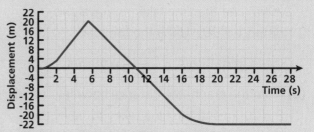

A Displacement–Time Graph for a Ball Kicked Against a Wall

Time	Distance Covered	Displacement
5s	18m	18m
11s	20m + 20m = 40m	20m – 20m = 0
22s	20m + 20m + 22m = 62m	20m – 20m – 22m = -22m

The gradient of a displacement–time graph is the velocity.

Calculating Speed

The speed of an object can be calculated by working out the gradient of a distance–time graph. The steeper the gradient, the faster the speed. All we have to do is take any point on the gradient and read off the distance travelled at this point, and the time taken to get there. Then use the formula:

$$\text{Speed} = \frac{\text{Distance travelled}}{\text{Time taken}}$$

Example

The graph below shows how the distance travelled by an object changes with time.

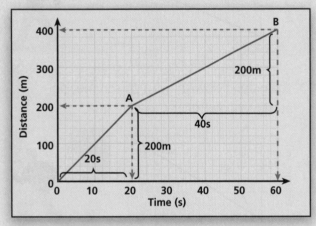

Find the speed of the object **(a)** from 0 to A, and **(b)** from A to B.

(a) From the graph, the distance from 0 to A is 200m and the time taken to travel this distance is 20s. Using the formula:

$$\text{Speed from 0 to A} = \frac{200m}{20s} = \textbf{10m/s}$$

(b) From the graph, the distance between A and B is 200m and the time taken to travel this distance is 60 – 20 = 40s. Using the formula:

$$\text{Speed from A to B} = \frac{200m}{40s} = \textbf{5m/s}$$

In this example, for the section A to B you need to find the *difference* in time and distance between points A and B. Do not just read the values from the axes. Using the values from the axes for point B will give you the *average* time for the whole journey from 0 to B, in this case 400m ÷ 60s = 6.7m/s.

This example illustrates how to calculate speed from a graph, but remember that this only works when looking at straight-line sections.

Curvy Distance–Time Graphs

The slope or gradient of a distance–time graph represents how quickly an object is moving; the steeper the slope, the faster its speed. When the gradient is changing, the object's speed is changing.

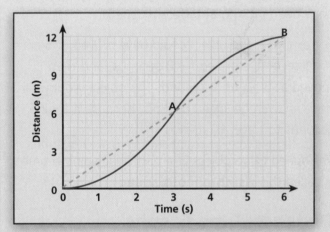

In section 0 to A the line is curved. This tells us that the object is speeding up because the gradient is increasing.

Between A and B the line curves the other way. The gradient is decreasing so the object must be slowing down.

Because the graph is curved it is difficult to work out the instantaneous speed at a particular point, but we can work out the average speed. The average speed is simply the total distance divided by the total time; it is shown by the dotted line.

$$\text{Speed} = \frac{\text{Distance travelled}}{\text{Time taken}}$$
$$= \frac{12m}{6s}$$
$$= \textbf{2m/s}$$

Where the gradient is steeper than the dotted line, the object is travelling faster than the average speed. Where the gradient is less steep than the dotted line, the object is travelling slower than the average speed.

Acceleration

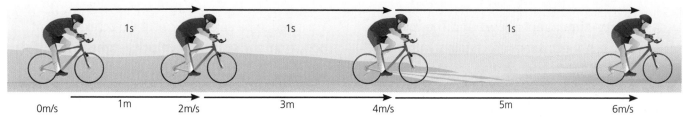

| 0m/s | 1m | 2m/s | 3m | 4m/s | 5m | 6m/s |

The **acceleration** of an object is the rate at which its velocity changes. In other words, it is a measure of how quickly an object speeds up or slows down.

The cyclist above increases his velocity by 2 metres per second every second. So, we can say that his acceleration is 2m/s² (2 metres per second, per second). To work out the acceleration of any moving object you need to know two things:

- The change in velocity.
- The time taken for this change in velocity.

You can then calculate the acceleration of the object using the following formula:

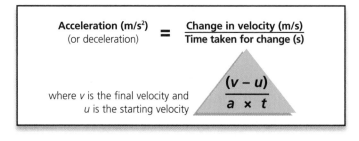

$$\textbf{Acceleration (m/s}^2\textbf{)} \text{ (or deceleration)} = \frac{\textbf{Change in velocity (m/s)}}{\textbf{Time taken for change (s)}}$$

$$\frac{(v-u)}{a \times t}$$

where *v* is the final velocity and *u* is the starting velocity

There are two important points to be aware of:

1. The cyclist above increases his velocity by the *same amount* every second; the *actual* distance travelled each second increases.

2. Deceleration is simply a negative acceleration, i.e. it describes an object that is slowing down.

Example

A cyclist accelerates uniformly from rest and reaches a velocity of 10m/s after 5s, then decelerates uniformly and comes to a halt in a further 10s. Calculate his acceleration **(a)** in the first 5s, and **(b)** as he is slowing down in the next 10s.

(a) $\text{Acceleration} = \dfrac{\textbf{Change in velocity}}{\textbf{Time taken}}$

$$= \frac{10-0}{5} = \textbf{2m/s}^2$$

(b) $\text{Acceleration} = \dfrac{\textbf{Change in velocity}}{\textbf{Time taken}}$

$$= \frac{0-10}{10} = \textbf{-1m/s}^2$$

The minus sign shows this is a deceleration, so the deceleration is 1m/s².

Speed–Time Graphs

The slope of a **speed–time graph** represents the acceleration of the object; the steeper the slope, the greater the acceleration.

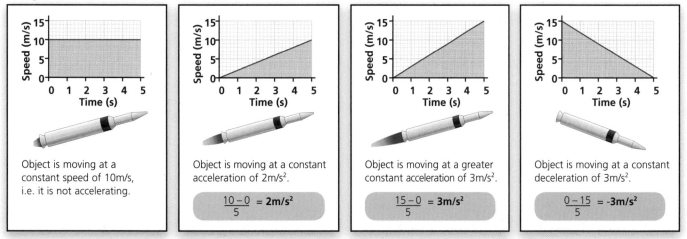

Object is moving at a constant speed of 10m/s, i.e. it is not accelerating.

Object is moving at a constant acceleration of 2m/s².

$$\frac{10-0}{5} = \textbf{2m/s}^2$$

Object is moving at a greater constant acceleration of 3m/s².

$$\frac{15-0}{5} = \textbf{3m/s}^2$$

Object is moving at a constant deceleration of 3m/s².

$$\frac{0-15}{5} = \textbf{-3m/s}^2$$

Velocity–Time Graphs

A **velocity–time** graph shows how the velocity at which an object is moving changes with time. Velocity has a direction, so if moving in a straight line in one direction is a positive velocity, then moving in a straight line in the opposite direction will be a negative velocity.

The graph below is the velocity–time graph for someone swimming lengths in a swimming pool.

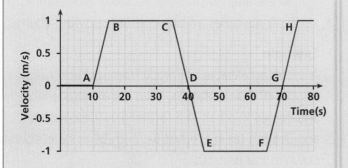

From 0 to A the swimmer is stationary – the velocity is zero. This is shown by a horizontal line along the time axis at zero velocity.

From A to B the swimmer is accelerating up the pool, and reaches a velocity of +1m/s after 15s.

Section B to C is a horizontal line at +1m/s. This shows that the swimmer is moving up the pool with constant velocity.

Section C to D shows the swimmer decelerating as they reach the end of the pool, turning round at D, where the velocity is zero once again.

Section D to E shows the swimmer accelerating to reach a velocity of -1m/s. The swimmer is now swimming at a constant velocity, but in the opposite direction, down the pool (section E to F).

Section F to G shows the swimmer decelerating to 0m/s at G, where they turn round. In section G to H the swimmer is once again accelerating up the pool, to reach a velocity of +1m/s.

Calculating Acceleration

You need to be able to calculate the acceleration of a body from the slope of a **velocity–time graph**.

Example

Here is a velocity–time graph. Calculate the acceleration of the three parts of the journey using the following formula:

$$\text{Acceleration} = \frac{\text{Change in velocity}}{\text{Time taken}}$$

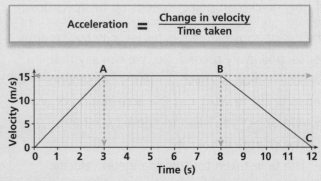

0 to A

From the graph, the time taken to reach a velocity of 15m/s is 3s.

Acceleration from 0 to A = $\frac{15\text{m/s}}{3\text{s}}$ = **5m/s²**

A to B

From A to B the line on the graph is horizontal, so the acceleration is zero.

Acceleration from A to B = $\frac{0\text{m/s}}{5\text{s}}$ = **0m/s²**

B to C

From B to C the velocity decreases from 15m/s to 0m/s in a time of 4s.

> The minus sign shows the object is slowing down, so this is a deceleration.

Acceleration from B to C = $\frac{-15\text{m/s}}{4\text{s}}$ = **-3.75m/s²**

So, the object accelerated at 5m/s² for 3 seconds, travelled at a constant speed of 15m/s for 5 seconds before decelerating at a rate of 3.75m/s² for 4 seconds.

Forces

A force occurs when two objects interact with each other. Whenever one object exerts a force on another, it always experiences an equal and opposite force in return. The forces in an **interaction pair** are equal in size and opposite in direction.

Examples

- **Gravity** (**weight**) – two masses are attracted to each other, e.g. we are attracted to the Earth and the Earth is attracted to us with an **equal** and **opposite** force. We do not notice the attraction of the Earth to us because it has such a big mass that our force has very little effect on it.
- If two people are standing on skateboards and one pushes the other, both skateboards will move away from each other.

- The rocket's engines push gas backwards (action) and the gas pushes the rocket forwards (reaction), thrusting it through the atmosphere. A jet engine works in a similar way.

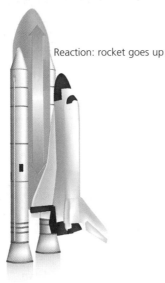

Reaction: rocket goes up

Action: gas rushes down

Friction and Reaction

Some forces, such as friction and reaction (of a surface), only occur as a response to another force.

A force occurs when an object is resting on a surface. The object is being pulled down to the surface by gravity and the surface pushes up with an equal force called the **reaction of the surface**.

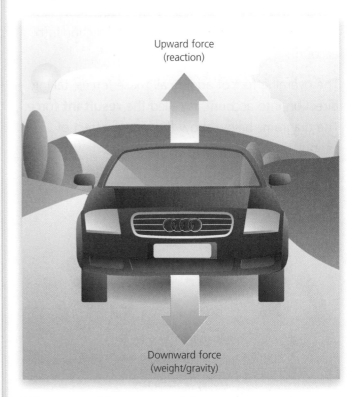

Upward force
(reaction)

Downward force
(weight/gravity)

When two objects try to slide past one another, both objects experience a force that tries to stop them moving. This interaction is called **friction**.

A moving object experiences friction. However, an object does not have to be moving to experience friction. For example, a car parked on a slope is trying to roll down the hill, but there is enough friction from the brakes (hopefully) to stop it.

Forces and Motion

Arrows are used when drawing diagrams of forces. The size and direction of the arrow represents the size of the force and the direction that it is acting in. Force arrows are always drawn with the tail of the arrow touching the object, even if the force is pushing the object.

If more than one force acts on an object, the forces need to be added, taking the direction into account.

The overall effect of adding all these forces, taking direction into account, is called the **resultant** force. The diagrams below show the overall effect of various push and friction forces on the movement of a trolley.

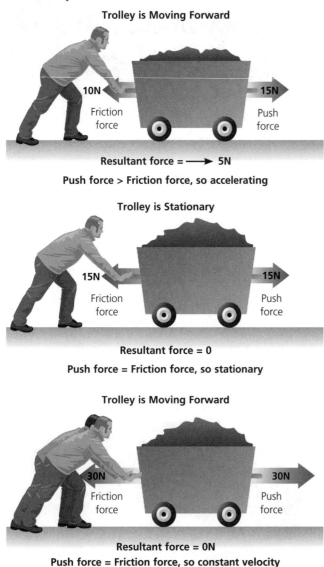

Trolley is Moving Forward

10N — Friction force

15N — Push force

Resultant force = ➞ 5N

Push force > Friction force, so accelerating

Trolley is Stationary

15N — Friction force

15N — Push force

Resultant force = 0

Push force = Friction force, so stationary

Trolley is Moving Forward

30N — Friction force

30N — Push force

Resultant force = 0N

Push force = Friction force, so constant velocity

Speeding Up and Slowing Down

A person walking, cars and bicycles have a **driving force**. In the case of a car, the driving force is produced by the engine. The person's muscles provide the driving force for someone walking or cycling.

However, there is also a **counter force**, caused by friction and air resistance, which has the effect of making the vehicle slow down or stop.

If the driving force is bigger than the counter force the vehicle accelerates.

100N ◄——— Car ———► 500N

Counter force 100N ◄——— Car ———► Driving force 500N

Car speeds up

If the driving force and counter force are equal, the vehicle travels at a constant speed in a straight line.

500N ◄——— Car ———► 500N

Counter force 500N ◄——— Car ———► Driving force 500N

Car travels at a constant speed

If the counter force is bigger than the driving force, the vehicle decelerates.

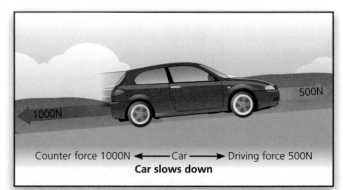

1000N ◄——— Car ———► 500N

Counter force 1000N ◄——— Car ———► Driving force 500N

Car slows down

Terminal Velocity

Falling objects experience two forces:

- the downward force of weight, W (↓) which always stays the same
- the upward force of air resistance, R, or drag (↑).

When a skydiver jumps out of an aeroplane, the speed of his descent can be considered in two separate parts – before and after the parachute opens:

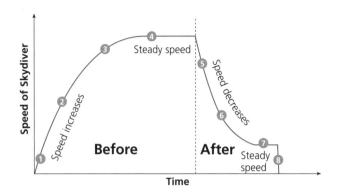

Before the Parachute Opens

When the skydiver jumps, he initially accelerates due to the force of gravity (see ❶). Gravity is a force of attraction that acts between objects that have mass, e.g. the skydiver and the Earth. The weight of an object is the force exerted on it by gravity. It is measured in newtons (N).

However, as the skydiver falls he experiences the frictional force of air resistance (R) in the opposite direction. But this is not as great as W so he continues to accelerate (see ❷).

As his speed increases, so does the air resistance acting on him (see ❸), until eventually R is equal to W (see ❹). This means that the resultant force acting on him is now zero and his falling speed becomes constant. This speed is called the **terminal velocity**.

After the Parachute Opens

When the parachute is opened, unbalanced forces act again because the upward force of R is now greatly increased and is bigger than W (see ❺). This decreases his speed, and as his speed decreases, so does R (see ❻).

Eventually R decreases until it is equal to W (see ❼). The forces acting are once again balanced, and for the second time he falls at a steady speed, slower than before though, i.e. at a new terminal velocity, until he lands on the ground (at ❽).

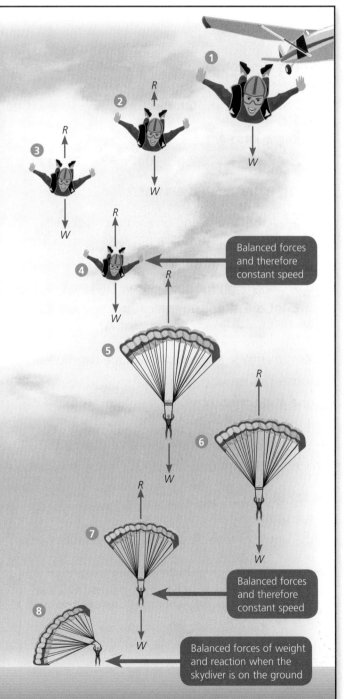

Momentum

Momentum is a measure of the motion of an object. The momentum of an object is calculated using the following formula:

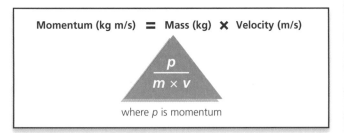

Momentum (kg m/s) **=** Mass (kg) **✕** Velocity (m/s)

$$\frac{p}{m \times v}$$

where *p* is momentum

Example

A car has a mass of 1200kg. It is travelling at a velocity of 30m/s. Calculate its momentum.

Using the formula:

Momentum = Mass × Velocity

= 1200kg × 30m/s **= 36 000kg m/s**

Change in Momentum

If the resultant force acting on an object is zero its momentum will not change. If the object:

- is stationary, it will remain stationary
- is already moving, it will continue moving in a straight line at a steady speed.

If the resultant force acting on an object is not zero, it causes a change in momentum in the direction of the force. This could:

- give a stationary object momentum (i.e. make it move)
- increase or decrease the speed, or change the direction, of a moving object, i.e. change its velocity.

The size of the change in momentum depends on:

- the size of the resultant force
- the length of time the force is acting on the object.

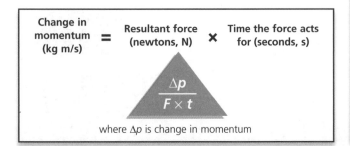

| Change in momentum (kg m/s) | **=** | Resultant force (newtons, N) | **✕** | Time the force acts for (seconds, s) |

$$\frac{\Delta p}{F \times t}$$

where Δ*p* is change in momentum

Collisions

If a car is involved in a collision it comes to a sudden stop, i.e. it undergoes a change in momentum. If this change in momentum is spread out over a longer period of time, the average resultant force acting on the car will be smaller.

Example

A car with a mass of 1000kg, travelling at 10m/s, has a momentum of 10 000kg m/s. If the car is involved in a collision and comes to a sudden stop, it would experience a change in momentum of 10 000kg m/s.

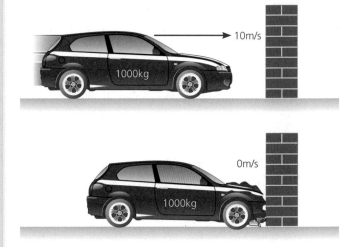

This sudden change in momentum will affect not only the car, but its passengers. If the car stops in a very short time, then the occupants of the car will experience a large force that can lead to serious injuries.

If the car stops in 0.05 seconds the force acting on it would be 200 000N, but if the car stops in 0.10 seconds the force would be 100 000N.

All car safety devices, such as seat-belts, crumple zones and air bags, are designed to reduce the force of the impact on the human body. They do this by increasing the time of the impact.

For example, a crumple zone is an area designed to 'crumple' on impact. This helps to increase the time during which the car changes momentum, i.e. instead of coming to an immediate halt there will be a longer time during which the momentum is reduced. This means that the force exerted on the people inside the car will be reduced, resulting in fewer injuries.

Kinetic Energy

A moving object has **kinetic** energy. The amount of kinetic energy an object has depends on:
- the mass of the object
- the velocity of the object.

The greater the mass and velocity of an object, the more kinetic energy it has. Therefore, it takes a longer time to stop a heavy, fast-moving object, than a light, slow-moving object.

Car brakes convert the kinetic energy into heat energy. As with all energy transfers, the total amount of energy remains the same (energy is conserved).

Kinetic energy is calculated using the following formula:

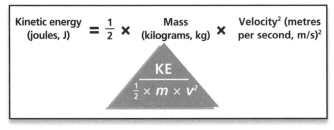

$$\text{Kinetic energy (joules, J)} = \frac{1}{2} \times \text{Mass (kilograms, kg)} \times \text{Velocity}^2 \text{ (metres per second, m/s)}^2$$

$$\frac{KE}{\frac{1}{2} \times m \times v^2}$$

Example

A bicycle and rider have a mass of 100kg and are moving at a velocity of 8m/s. How much kinetic energy do they have?

8m/s

100kg

Using the formula:

$$\text{Kinetic energy} = \frac{1}{2} \times \text{Mass} \times \text{Velocity}^2$$

$$= \frac{1}{2} \times 100\text{kg} \times (8\text{m/s})^2$$

$$= \frac{1}{2} \times 100 \times 64$$

$$= \textbf{3200J}$$

You need to be able to discuss the transformation of kinetic energy to other forms of energy in particular situations.

Example: Space Shuttles

When a space shuttle returns to Earth it has a lot of kinetic energy, i.e. it has a large mass and is travelling fast. As it enters the Earth's atmosphere, the shuttle encounters frictional forces and the kinetic energy is transformed into heat energy, which slows it down.

The shuttle can reach extremely high temperatures because of the heat energy produced, which causes a risk of fire and explosion. Scientists have developed special heat shields to try to protect the body of space shuttles (and the astronauts in them) from this intense heat.

Gravitational Potential Energy

An object lifted above the ground gains potential energy (PE), often called **gravitational potential energy** (**GPE**). The additional height gives it the potential to do work when it falls, e.g. a diver standing on a diving board has gravitational potential energy. It is calculated by the following formula:

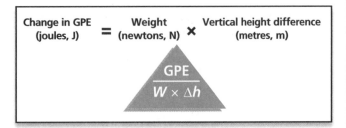

If an object is dropped, its gravitational potential energy decreases and is converted into kinetic energy.

Example

On Earth, an object of mass 2kg weighs 20N. How much kinetic energy does the object gain if it is dropped from a height of 5m?

N.B. Remember, to find the GPE you use the weight (20N), not the mass.

Change in GPE = Weight × Height difference
= 20N × 5m = **100J**

If the object loses 100J of gravitational potential energy when it falls, it must have gained 100J of kinetic energy.

HT   To work out the velocity of a falling object you need to use the mass. In the example above, we know that the object has gained 100J of kinetic energy.

Kinetic energy = $\frac{1}{2}$ × Mass × Velocity²

$100 = \frac{1}{2} \times 2 \times velocity^2$

$100 = 1 \times velocity^2$

$100 = velocity^2$

$velocity = \sqrt{100}$

= **10m/s**

Work and Energy

When a force moves an object, work is done on the object resulting in the transfer of energy.

When work is done *by* an object it loses energy, and when work is done *on* an object it gains energy according to the relationship:

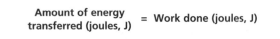

When a force acting on an object causes it to accelerate, work is done on the object to increase its kinetic energy. A falling object has work done on it by the force of gravity and this causes it to accelerate.

If we ignore the effects of air resistance and friction, the increase in kinetic energy will be equal to the amount of work done. However, in reality some of the energy will be dissipated (lost) as heat, and the increase in kinetic energy will, therefore, be less than the work done.

When a car is travelling at a constant speed the driving force from the engine is balanced by the counter force. All of the work being done by the engine is used to overcome friction and there is no increase in kinetic energy.

Work done is calculated by the following formula:

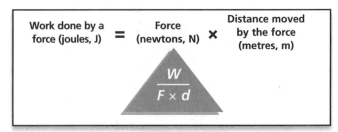

Example

A car needs to be pushed with a force of 100N to overcome friction. If the car is pushed for 5m, calculate the work done.

Work done = Force × Distance moved
= 100N × 5m
= **500J**

Module P5 (Electric Circuits)

When electrical charges move, they create a current. Different types of current have different uses. This module looks at:
- static electricity
- electric currents
- electromagnetic induction
- electrical generators
- the use of transformers.

Static Electricity

Some insulating materials can become electrically charged when they are rubbed against each other. The electrical charge then stays on the material, i.e. it does not move (the charge is 'static').

Charge builds up when electrons (which have a negative charge) are 'rubbed off' one material onto another. The material receiving the electrons becomes negatively charged and the one giving up electrons becomes positively charged.

If a Perspex rod is rubbed with a cloth, the rod loses electrons to become positively charged. The cloth gains electrons to become negatively charged.

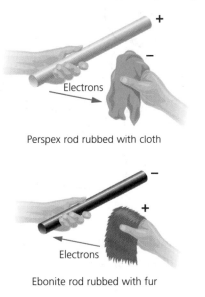

Perspex rod rubbed with cloth

Ebonite rod rubbed with fur

If an ebonite rod is rubbed with fur, the ebonite gains electrons to become negatively charged. The fur loses electrons to become positively charged.

Repulsion and Attraction

When two charged materials are brought together, they exert a force on each other so they are attracted or repelled. Two materials with the same type of charge repel each other; two materials with different charges attract each other.

If a positively charged Perspex rod is moved near to another positively charged Perspex rod suspended on a string, the suspended rod will be repelled. We would get the same result with two negatively charged ebonite rods.

If a negatively charged ebonite rod is moved near to a positively charged suspended Perspex rod, the suspended Perspex rod will be attracted. We would get the same result if the rods were the other way round.

Example

When cars are spray painted, a panel of the car is positively charged and the paint is negatively charged. The paint particles repel each other, but are attracted to the positively charged panel. This causes the paint to be applied evenly.

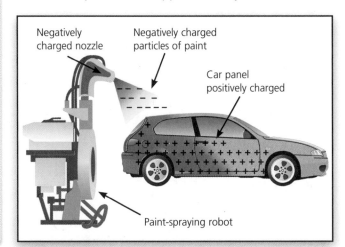

Negatively charged nozzle

Negatively charged particles of paint

Car panel positively charged

Paint-spraying robot

Electric Circuits

An **electric current** is a **flow of charge**. It is measured in **amperes** (amps).

In an electric circuit the components and wires are full of charges that are free to move. When a circuit is made, the battery causes these charges to move in a continuous loop. The charges are not used up.

In metal **conductors** there are lots of charges free to move. **Insulators**, on the other hand, have few charges that are free to move. Remember, metals contain free electrons in their structure. The movement of these electrons creates the flow of charge (electric current).

Direction of current

Circuit Symbols

You need to know the following standard symbols:

Cell	—|⊢—
Battery of cells	—|⫴⊢— or —|⊢··|⊢—
Power supply	—o o—
Filament lamp	—⊗—
Switch (open)	—⟋ o—
Switch (closed)	—o‒o—
Light dependent resistor (LDR)	—▱—
Fixed resistor	—▭—
Variable resistor	—▱⟋—
Thermistor	—▱—
Voltmeter	—Ⓥ—
Ammeter	—Ⓐ—

Current Size

The amount of current flowing in a circuit depends on the resistance of the components in the circuit and the **potential difference** (p.d.) across them. Potential difference tells us the energy given to the charge and is another name for **voltage**. It is a measure of the 'push' of the battery on the charges in the circuit.

For example, a 12 volt battery will transfer 12 joules of energy to every unit of charge. A potential difference of 3 volts across a bulb means that the bulb is transferring 3 joules of energy from every unit of charge; this energy is transferred as heat and light.

The greater the potential difference (or voltage) across a component, the greater the current that flows through the component. Two cells together provide a bigger potential difference across a lamp than one cell. This makes a bigger current flow.

Current is measured using an ammeter and voltage (or potential difference) is measured using a voltmeter (see page 82).

Resistance and Current

Components such as resistors, lamps and motors resist the **flow** of **charge** through them, i.e. they have **resistance**. Work is done by the power supply and energy is transferred to the component.

The greater the resistance of a component or components, the smaller the current that flows for a particular voltage, or the greater the voltage needed to maintain a particular current.

Even the connecting wires in the circuit have some resistance, but it is such a small amount that it is usually ignored.

Two lamps together in a circuit with one cell have a certain resistance. Including another cell in the circuit provides a greater potential difference, so a larger current flows.

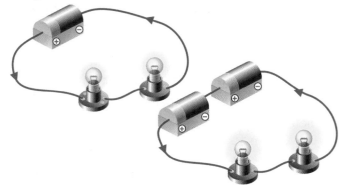

Adding resistors in series increases the resistance because the battery has to push charges through all of the resistors.

Adding resistors in parallel reduces the total resistance and increases the total current because this provides more paths for the charges to flow along.

When an electric current flows through a component it causes the component to heat up. In a filament lamp this heating effect is large enough to make the filament in the lamp glow.

> **HT** As the current flows, the moving charges collide with the vibrating ions in the wire giving them energy; this increase in energy causes the component to become hot.

Resistance, which is measured in ohms, Ω, is a measure of how hard it is to get a current through a component at a particular potential difference or voltage. Resistance can be calculated by the following formula:

$$\text{Resistance (ohms, } \Omega\text{)} = \frac{\text{Voltage (volts, V)}}{\text{Current (amperes, A)}}$$

where I is current

$$\frac{V}{I \times R}$$

Example

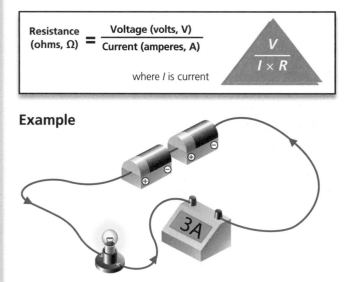

The circuit above has a current of 3 amps and a voltage of 6V. What is the resistance?

$$\text{Resistance} = \frac{\text{Voltage}}{\text{Current}} = \frac{6V}{3A} = \mathbf{2\Omega}$$

> **HT** The resistance formula can be rearranged using the formula triangle to work out the voltage or current.
>
> ### Example
> Calculate the reading on the voltmeter in this circuit if the bulb has a resistance of 15 ohms.
>
>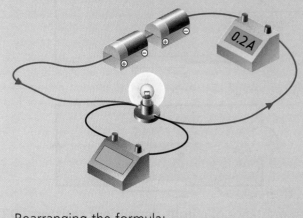
>
> Rearranging the formula:
> Voltage = Current × Resistance
> = 0.2A × 15Ω = **3V**

Current–Voltage Graphs

As long as the temperature of a resistor stays constant, the current through the resistor is directly proportional to the voltage across the resistor, regardless of which direction the current is flowing, i.e. if one doubles, the other also doubles.

This means that a graph showing current through the component and voltage across the component will be a straight line through 0. (If there is no voltage, then there isn't anything to push the current around.)

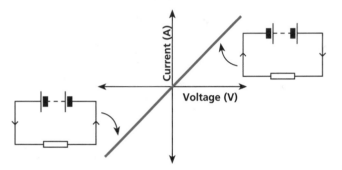

Thermistors and Light Dependent Resistors (LDRs)

The resistance of some materials depends on environmental conditions.

The resistance of a **thermistor** depends on its temperature. Its resistance decreases as the temperature increases; this allows more current to flow.

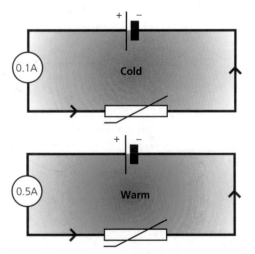

The resistance of a **light dependent resistor** (**LDR**) depends on light intensity. Its resistance decreases as the amount of light falling on it increases; this allows more current to flow.

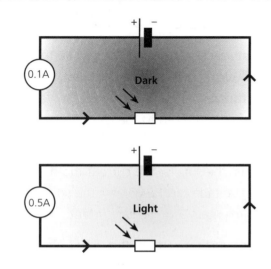

As the resistance of an LDR and a thermistor can change, this will result in a change in the potential difference for all the other components in the circuit.

Potential Difference and Current

The potential difference across a component in a circuit is measured in volts (V) using a voltmeter connected in parallel across the component.

The current flowing through a component in a circuit is measured in amperes (A), using an ammeter connected in series.

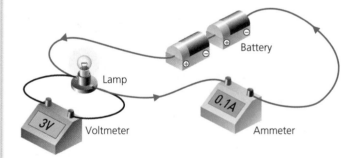

When batteries are added in series, the total potential difference is the sum of all the individual potential differences.

> **HT** When batteries are added in parallel, the total potential difference and current through the circuit remains the same, but each battery supplies less current.
>
> This sharing of the load makes the batteries last longer.

Series Circuits

In a series circuit, all components are connected one after another in one loop going from one terminal of the battery to the other.

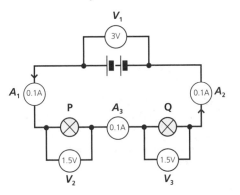

- The current flowing through each component is the same, i.e. $A_1 = A_2 = A_3$.
- The potential difference across the components adds up to the potential difference across the battery, i.e. $V_1 = V_2 + V_3$.

> **HT** The total energy transferred to each unit charge by the battery must equal the total amount of energy transferred from the charge by the component because energy cannot be created or destroyed.

- If another battery was added to the above circuit, it would have the effect of increasing the voltage and current.
- The potential difference is largest across components with the greatest resistance.

> **HT** More energy is transferred from the charge flowing through a greater resistance because it takes more energy to push the current through the resistor.
>
> If a circuit has several identical components, they will have the same potential difference across them. If the components are different, the one with the greatest resistance will have the greatest potential difference across it, and vice versa.

Parallel Circuits

Components connected in parallel are connected separately in their own loop going from one terminal of the battery to the other.

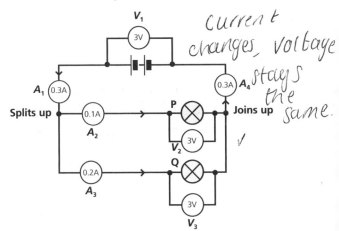

current changes, voltage stays the same. ✓

- The amount of current that passes through each component depends on the resistance of each component. The greater the resistance, the smaller the current. In the circuit above, bulb Q has half the resistance of bulb P, so twice as much current flows through it.
- The total current from the battery is equal to the sum of the current through each of the parallel components, i.e. $A_1 = A_2 + A_3 = A_4$. In the circuit above, 0.3A = 0.1A + 0.2A = 0.3A.
- The current is smallest through the component with the greatest resistance.

> **HT** The same voltage causes more current to flow through a smaller resistance than a bigger one.
>
> The potential difference across each component is equal to the potential difference of the battery.
>
> The current through each component is the same as if it was the only component present. For example, a circuit with a battery and a bulb has a 1 amp current. If a second identical component is added in parallel, the current through each would be 1 amp. The total current through the battery will increase.

Electromagnetic Induction

In electromagnetic induction, **movement produces voltage**. If a wire or a coil of wire cuts through the lines of force of a magnetic field (or vice versa), then a voltage is induced (produced) between the ends of the wire. If the wire is part of a complete circuit, a current will be induced.

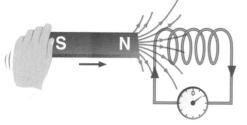

Moving the magnet into the coil induces a current in one direction. A current can be induced in the opposite direction by:

* moving the magnet out of the coil

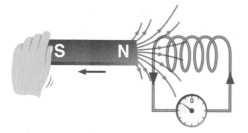

* moving the other pole of the magnet into the coil.

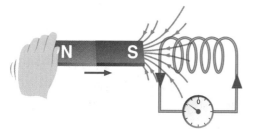

Both of these involve a magnetic field being cut by a coil of wire, creating an induced voltage. If there is no movement of the magnet or coil there is no induced current, because the magnetic field is not being cut.

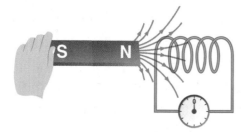

The Electric Generator

Mains electricity is produced by generators. Generators use the principle of electromagnetic induction to generate electricity by rotating a magnet inside a coil.

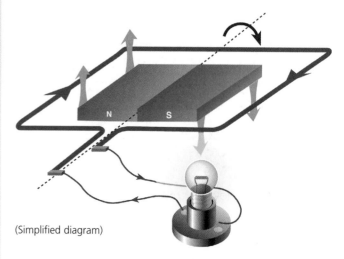

(Simplified diagram)

The size of the induced voltage follows the same rules as with the magnet and coil. The voltage will be increased by:

* increasing the speed of rotation of the magnet
* increasing the strength of the magnetic field
* increasing the number of turns on the coil
* placing an iron core inside the coil.

HT As the magnet rotates, the **voltage** induced in the coil changes direction and size, as seen below.

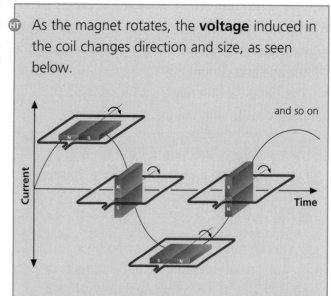

The induced current reverses its direction of flow every half turn of the magnet and is, therefore, alternating current.

Power

When charge flows through a component, energy is transferred to the component. Power, measured in watts (W), is a measure of how much energy is transferred every second, i.e. **the rate of energy transfer**.

Electrical power can be calculated using the following equation:

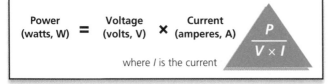

Power (watts, W)	=	Voltage (volts, V)	×	Current (amperes, A)

$$\frac{P}{V \times I}$$

where *I* is the current

Example
An electric motor works at a current of 3A and a potential difference of 24V. What is the power of the motor?

Power = Potential difference × Current
= 24V × 3A
= **72W**

HT The power formula can be rearranged using the formula triangle to work out the potential difference or current.

Example
A 4W light bulb works at a current of 2A. What is the potential difference?

$$\text{Potential difference} = \frac{\text{Power}}{\text{Current}}$$
$$= \frac{4W}{2A}$$
$$= \textbf{2V}$$

Transformers

Primary coil — Iron core — Secondary coil

Transformers are used to change the voltage of an alternating current. They consist of two coils, called the **primary** and **secondary** coils, wrapped around a soft iron core.

HT When two coils of wire are close to each other, a changing magnetic field in one coil can induce a voltage in the other. Alternating current flowing through the primary coil creates an alternating magnetic field. This changing field then induces an alternating voltage across the secondary coil, which causes an alternating current through the secondary coil.

The amount by which a transformer changes the voltage depends on the number of turns on the primary and secondary coils. If the number of turns on each coil is the same, the voltage does not change. If there are more turns on the secondary coil, the voltage goes up; if there are fewer turns, the voltage goes down.

You need to be able to use this equation:

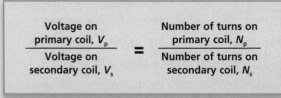

$$\frac{\text{Voltage on primary coil, } V_p}{\text{Voltage on secondary coil, } V_s} = \frac{\text{Number of turns on primary coil, } N_p}{\text{Number of turns on secondary coil, } N_s}$$

Example
A transformer has 1000 turns on the primary coil and 200 turns on the secondary coil.

If a voltage of 250V is applied to the primary coil, what is the voltage across the secondary coil?

Using our formula: $\dfrac{V_p}{V_s} = \dfrac{N_p}{N_s}$

$$\frac{250}{V_s} = \frac{1000}{200}$$
$$250 = 5V_s$$
$$V_s = \frac{250}{5}$$
$$V_s = \textbf{50V}$$

How an a.c. Generator Works

In simple terms, an a.c. generator works by placing a coil of wire within a magnetic field and **rotating** it.

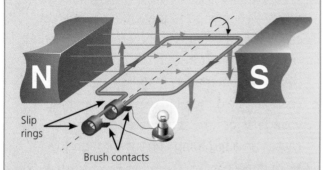

Slip rings

Brush contacts

As it rotates and cuts through the magnetic field, a voltage is induced across the coil and a current is induced in the coil. One side of the coil moves up during one half turn, and then down during the next half. This means that the current reverses direction, or **alternates**, every half turn (after it has turned through 180°).

Each end of the coil is attached to a separate metal ring, called a **slip ring**, which rotates with it. The generator can be connected to an external circuit (e.g. to power a light) using brush contacts. These brushes are spring-loaded so that they continuously push against the slip rings and the circuit remains complete. They need to be replaced regularly because they gradually wear away.

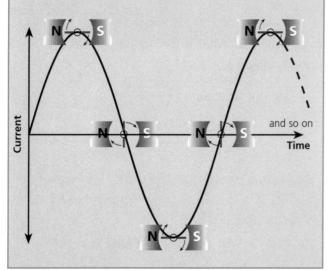

Types of Current

A **direct current** (d.c.) always flows in the same direction. Cells and batteries supply direct current.

An **alternating current** (a.c.) changes the direction of flow back and forth continuously. The number of complete cycles per second is called the **frequency**, and for UK mains electricity this is 50 cycles per second (hertz, Hz).

In the UK, the mains supply has a voltage of about 230 volts which, if it is not used safely, can kill.

We use a.c. instead of d.c. for our electrical supply for different reasons:

- a.c. is easier to generate and distribute over large distances within the National Grid. When d.c. is transmitted over large distances, it loses a lot of electrical energy within the wires due to heat, etc. This would mean that the further you live from the power station, the less energy you would get.
- a.c. can be increased and decreased using transformers that reduce energy loss due to heat. Therefore a.c. can be transmitted with hardly any power loss. d.c. cannot be used with transformers. At the power station, transformers step up the voltage to between 150 000V and 400 000V. Before electricity is consumed by the domestic user, transformers step down the voltage of the electricity to a level that is safe to use, e.g. 230V.

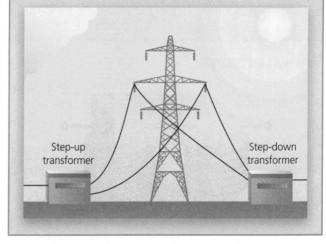

Step-up transformer

Step-down transformer

The Principles of the Motor Effect

In the motor effect, **current produces movement**. When a conductor (wire) carrying an electric current is placed in a magnetic field, the magnetic field formed around the wire interacts with the permanent magnetic field causing the wire to experience a force, which makes it move. This force acts at right angles to both the current and magnetic field. (The force is shown by green arrows in the diagrams below.)

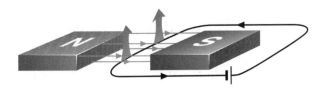

The **size** of the force on the wire can be increased by:

- increasing the size of the current (e.g. having more cells)

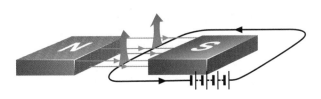

- increasing the strength of the magnetic field (e.g. having stronger magnets).

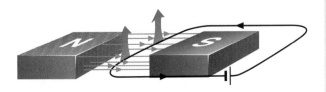

The **direction** of the force on the wire can be reversed by:

- reversing the direction of flow of the current (e.g. turning the cell around)

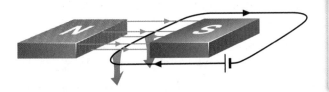

- reversing the direction of the magnetic field (e.g. swapping the magnets around).

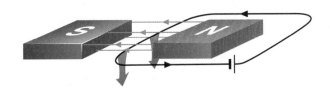

The wire will not experience a force if it is parallel to the magnetic field.

The Direct Current Motor

Electric motors rely on the principle of the motor effect. They form the basis of a vast range of electrical devices both inside and outside the home.

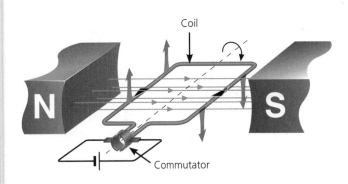

As a current flows through the coil, a magnetic field is formed around the coil, creating an electromagnet. This magnetic field interacts with the permanent magnetic field that exists between the two poles, North and South. A force acts on both sides of the coil, which rotates the coil to give us a very simple motor.

To make the motion continuous, a commutator is used. The commutator makes the direction of the current reverse every half turn, so the motor keeps turning in the same direction.

A motor can therefore make something move. Motors are used to move the drum in a washing machine, spin the disc in a DVD player and a hard disk drive, and turn the wheels of an electric car. Tiny electric motors drive the wheels in toy cars, and huge motors drive electric trains.

Module P6 (Radioactive Materials)

Some materials are radioactive because of the structure of the atoms they contain. Radiation is all around us and, while there are some dangers and hazards to address, certain radiations are very useful. This module looks at:

- the structure of the atom
- nuclear fusion
- alpha, beta and gamma radiation
- half-life
- background radiation
- uses of radiation
- nuclear fission and nuclear power stations.

Atoms and Elements

All elements are made of **atoms**; each element contains only one type of atom. All atoms contain a nucleus and electrons. The nucleus is made from protons and neutrons with the one exception of hydrogen (the lightest element), which has no neutrons – just one proton and one electron.

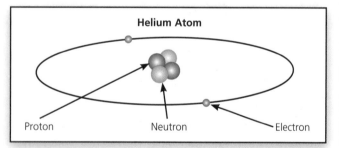

Helium Atom

Proton   Neutron   Electron

Some elements give out **ionising radiation** all the time; we call these elements **radioactive**. Neither chemical reactions nor physical processes (e.g. heating) can change the radioactive behaviour of a radioactive substance.

HT An atom has a nucleus made of protons and neutrons. Every atom of a particular **element** always has the same number of protons (if it contained a different number of protons it would be a different element). For example…

- hydrogen atoms have one proton
- helium atoms have two protons
- oxygen atoms have eight protons.

However, some atoms of the same element can have different numbers of neutrons – these are called **isotopes**. For example, these are three isotopes of oxygen:

| Oxygen-16 has eight neutrons | Oxygen-17 has nine neutrons | Oxygen-18 has ten neutrons |

All three of these isotopes have eight protons.

Ionising Radiation

There are three types of radiation that can be given out by radioactive materials. These radiations are **alpha**, **beta** and **gamma**.

Different radioactive materials will give out any one, or a combination, of these radiations.

An easy way to tell which type of radiation you are dealing with is to test its penetrating power.

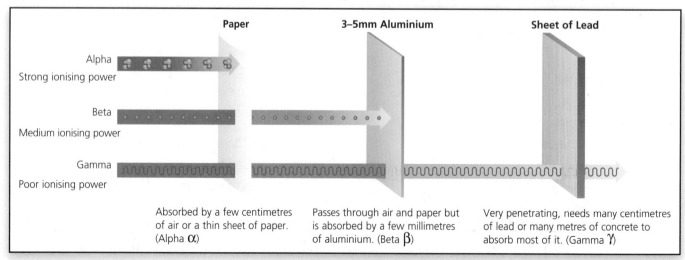

Paper   3–5mm Aluminium   Sheet of Lead

Alpha — Strong ionising power

Beta — Medium ionising power

Gamma — Poor ionising power

Absorbed by a few centimetres of air or a thin sheet of paper. (Alpha α)

Passes through air and paper but is absorbed by a few millimetres of aluminium. (Beta β)

Very penetrating, needs many centimetres of lead or many metres of concrete to absorb most of it. (Gamma γ)

Nuclear Fusion

At the beginning of the 20th century, discoveries about the nature of the atom and nuclear processes began to help answer the mystery of where the Sun's energy comes from.

In 1911, there was a ground-breaking experiment – the Rutherford–Geiger–Marsden alpha particle scattering experiment. In this experiment, a thin gold foil was bombarded with alpha particles. The effect on the alpha particles was recorded, and these observations provided the evidence for our current understanding of atoms.

Most alpha particles were seen to **pass straight through** the gold foil. This would indicate that gold atoms were composed of large amounts of open space. However, some particles were **deflected** slightly and a few were even **deflected back** towards the source. This would indicate that the alpha particles passed close to something positively charged within the atom and were repelled by it.

These observations brought Rutherford and Marsden to conclude that:

- gold atoms, and therefore all atoms, consist of largely empty space with a small, positive region called the **nucleus**
- the nucleus is positively charged
- the electrons are arranged around the nucleus with a great deal of space between them.

> **HT** Rutherford's data allowed him to identify a small positive nucleus, but the limited data did not explain its structure. Later experiments revealed the presence of positive protons and neutral neutrons in the nucleus held together by the short-range **strong nuclear force**. Protons normally repel each other (because they have the same charge), but the strong nuclear force balances the repulsive electrostatic force between the protons.

The strong nuclear force between protons and neutrons can also cause hydrogen nuclei (or protons) that are close enough to each other to fuse into helium nuclei. This process releases large amounts of

energy and is known as **nuclear fusion**. It is the same process that occurs inside the Sun.

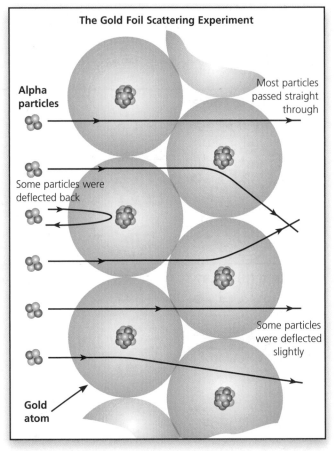

The Gold Foil Scattering Experiment

Alpha particles

Most particles passed straight through

Some particles were deflected back

Some particles were deflected slightly

Gold atom

HT Einstein's Equation

Perhaps one of Albert Einstein's greatest insights was to realise that matter and energy are really different forms of the same thing. He quantified this brilliantly with the famous equation:

$$E = mc^2$$

E = The energy produced
m = The mass lost
c = The speed of light (in a vacuum) 300 000 000 m/s

This same equation can be used to calculate the energy released during nuclear fusion and fission.

If two hydrogen nuclei are forced together, one of the protons will change into a neutron, resulting in the formation of a deuterium nucleus. The deuterium nucleus has a lower mass than the proton and neutron separately. This 'mass defect' is converted into energy and can be calculated using $E = mc^2$.

Radioactive Decay

The emission of ionising radiation occurs because the nucleus of an unstable atom is decaying. The type of decay depends on why the nucleus is unstable; the process of decay helps make the atom become more stable. During decay, the number of protons in the atom may change. If this happens, the element changes from one type to another.

Alpha (α) decay

Unstable parent ➡ New daughter + Alpha particle

$$^{238}_{92}\text{U} \rightarrow {}^{234}_{90}\text{Th} + {}^{4}_{2}\text{He}$$

The original atom decays by ejecting an alpha (α) particle from the nucleus. This particle is a helium nucleus: a particle made up of two protons and two neutrons. With alpha decay a new atom is formed. This new atom has two protons and two neutrons fewer than the original.

Beta (β) decay

Unstable parent ➡ New daughter + Beta particle

$$^{34}_{15}\text{P} \rightarrow {}^{34}_{16}\text{S} + {}^{0}_{-1}\text{e}$$

The original atom decays by changing a neutron into a proton and an electron. This high-energy electron, which is now ejected from the nucleus, is a beta (β) particle. With beta decay a new atom is formed. This new atom has one more proton and one less neutron than the original.

Gamma (γ) decay

Since a gamma ray has no mass, there is no change to the 'parent' element.

Unstable parent ➡ Stable parent + Gamma radiation

After α or β decay, a nucleus sometimes contains surplus energy. It emits this as gamma radiation, which is very high frequency electromagnetic radiation. During gamma decay, only energy is emitted. This does not change the type of atom.

Background Radiation

Radioactive elements are found naturally in the environment. The radiation produced by these sources contributes to the overall background radiation.

There is nothing we can do to prevent ourselves from being **irradiated** and **contaminated** by background radiation, but the level of background radiation in most places is so low that it is nothing to worry about. There is, however, a correlation between certain cancers and living for many years in areas where the underlying rock is granite.

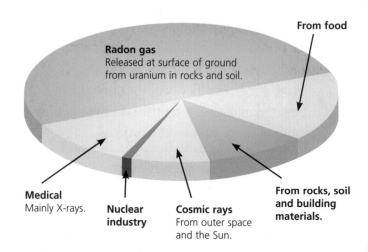

From food

Radon gas
Released at surface of ground from uranium in rocks and soil.

Medical
Mainly X-rays.

Nuclear industry

Cosmic rays
From outer space and the Sun.

From rocks, soil and building materials.

Half-life

The **activity** of a substance is a measure of the amount of radiation given out per second.

When a radioactive atom decays it becomes less radioactive and its activity drops. The **half-life** of a substance is the time it takes for the radiation emitted by a radioactive material to halve. The half-life of a radioactive material can range from a few seconds to millions of years.

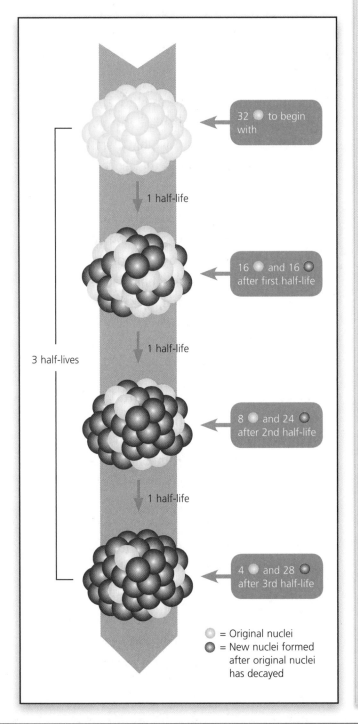

32 ● to begin with

1 half-life

16 ● and 16 ● after first half-life

1 half-life

3 half-lives

1 half-life

8 ● and 24 ● after 2nd half-life

1 half-life

4 ● and 28 ● after 3rd half-life

● = Original nuclei
● = New nuclei formed after original nuclei has decayed

Half-life and Safety

All radioactive substances become less radioactive as time passes. A substance would be considered safe once its activity had dropped to the same level emitted as background radiation, which is a dose of around 2 millisieverts per year or 25 counts per minute with a standard detector.

Some substances decay quickly and could be safe in a very short time. Substances with a long half-life remain harmful for millions of years.

HT Half-life Calculations

The half-life can be used to calculate how old a radioactive substance is, or how long it will take to become safe.

Example

If a sample has an activity of 800 counts per minute and a half-life of 2 hours, how many hours will it take for the activity to reach the background rate of 25 counts per minute?

We need to work out how many half-lives it takes for the sample of 800 counts to reach 25 counts.

1. $\frac{800}{2} = 400$

2. $\frac{400}{2} = 200$

3. $\frac{200}{2} = 100$

4. $\frac{100}{2} = 50$

5. $\frac{50}{2} = 25$

It takes 5 half-lives to reach a count of 25, and each half-life takes 2 hours, so it takes:

5×2 hours = **10 hours**

Ionisation

When radiation interacts with neutral atoms or molecules, the atoms or molecules may become charged due to electrons being knocked out of their structure. This alters their structure, leaving them as charged particles called **ions**. Alpha, beta and gamma radiation are therefore known as ionising radiation and can damage molecules in healthy cells, which results in the death of the cell.

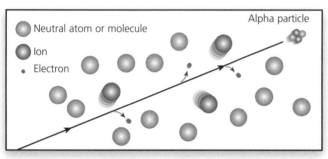

Cell Damage

When living cells absorb radiation, damage can occur in different ways:
- Ionising radiation can damage cells, causing ageing of the skin.
- Ionising radiation can cause mutations in the nucleus of a cell, which can lead to cancer.
- Different amounts of exposure can cause different effects, e.g. high-intensity ionising radiation can kill cells, leading to radiation poisoning.

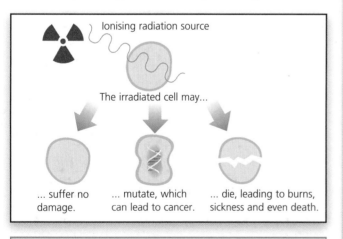

HT Ionising radiation is able to break molecules into bits (called ions), which can then take part in other chemical reactions.

Uses of Radiation

Although using ionising radiation can be dangerous, there are many beneficial uses:
- to treat cancer
- to sterilise surgical instruments and food
- as a tracer in the body for medical procedures.

HT Cancer Treatment

High-energy gamma rays can be used to kill cancer cells. However, ionising radiation can damage living cells too, so the radiation has to be carefully targeted from different angles to minimise the damage to healthy cells.

Radioactive iodine can be used to target thyroid cancer. Iodine is needed by the thyroid, so it collects naturally in the thyroid, where it gives out beta radiation and kills the cancer cells.

In both of these examples there is a danger of damage to healthy cells, so the doctors need to carefully weigh the risks against the benefits. When deciding on the dose of radiation to use, the **ALARA** (as low as reasonably achievable) principle should be applied. This means that measures should be taken to make the risks as small as possible, while still providing the benefits and taking into account all the implications.

Sterilising Surgical Instruments

Surgical instruments must not have any bacteria on them. Bacteria are living cells, so are susceptible to damage from ionising radiation. Exposing surgical instruments in sealed bags to gamma radiation results in the death of the bacterial cells, and therefore sterilisation of the surgical instruments. The instruments remain sterile until they are ready to be used.

Sterilising Food

Irradiating food kills any bacteria that would cause the food to go off. Radiation treatment is only allowed on a few foods in the UK and these have to carry a label stating that they have been treated with radiation.

Tracers in the Body

Doctors use radioactive chemicals called tracers to help detect damage to the internal organs. Once the tracer enters the body, it builds up in the damaged or diseased part of the body. Radiation detectors can then be used to detect where the problem is. These can be linked to computers, which produce an image showing the distribution of the radioactive chemical. Problem areas are highlighted by a high concentration.

Because it is being used inside the body, the radioactive tracer must be non-toxic. It also needs to have a short half-life, so that it breaks down quickly after use. Gamma and beta sources are used because they pass out of the body easily. An alpha source is never used because it would be quickly absorbed and cause damage.

Technetium-99 is a gamma emitter often used for this purpose. It has a half-life of 6 hours. This gives doctors enough time to detect the problem, but ensures that the radiation decreases to a safe level quickly.

Dangers of Radiation

New scientific advances often create an element of risk. The transportation and application of radioactive substances is carefully controlled by government rules and regulations to minimise the risk to the general public. However, people working in the nuclear industry, medical physics (X-rays, etc.) and many other areas often have to work with radioactive materials. They can become irradiated or contaminated by the materials, which could lead to serious health problems or even death. Therefore, the exposure that these people are subjected to needs to be carefully monitored.

The different types of radiation carry different risks:

- Alpha is the most dangerous if the source is inside the body; all the radiation will be absorbed by cells in the body.
- Beta is the most dangerous if the source is outside the body because, unlike alpha, it can penetrate the outer layer of skin and damage the internal organs.
- Gamma can cause harm if it is absorbed by the cells in the body, but it is weakly ionising and can pass straight through the body, causing little damage.

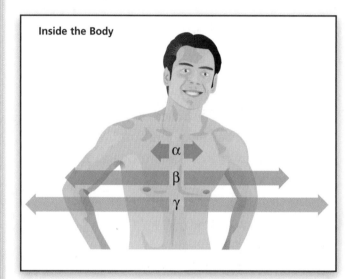

Inside the Body

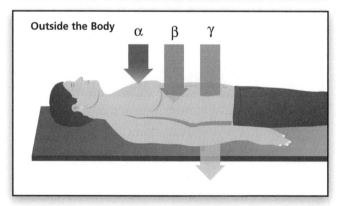

Outside the Body

The sievert is a measure of a radiation dose's potential to do harm to a person. It is based on both the type and amount of radiation absorbed.

A 5 sievert dose is a 5 sievert dose regardless of the type of radiation absorbed. This makes the sievert a very useful unit for radiation safety measures.

Radiation Protection

The transportation and application of radioactive substances is carefully controlled by government rules and regulations to minimise the risks to the general public. Authorised personnel who handle radioactive substances or operate machinery follow strict guidelines. These vary depending on the risk factor involved. The table below lists some measures that can be taken to reduce/prevent exposure to different types of radiation.

External Exposure – Beta and Gamma

- Minimise time of exposure.
- Maximise distance from the source of radiation.
- Wear protective clothing (and remove it before leaving a restricted area).
- Avoid direct handling of radioactive materials, i.e. use implements like forceps and tongs.
- Use protective shields, screens and containers.
- Use instruments that can detect levels of radiation, e.g. Geiger-Müller counter or liquid scintillation counter.
- Use materials that can provide a shield against radiation, e.g. lead, concrete and water.
- Wear a film badge that monitors the degree of exposure to radiation.

Internal Exposure – Alpha, Beta and Gamma

- Wear chemical fume hoods and protective masks to prevent inhalation.
- Never consume or store food and drink close to a radioactive source.
- Wear protective clothing.
- Ensure that any cuts or wounds are sealed up.
- Minimise amount of radioactive material to be handled.

People who work in the nuclear industry, such as radiographers and nuclear power plant technicians, are regularly exposed to radiation. They often wear a badge to monitor their degree of exposure. The badge contains photographic film, which (after developing) becomes darker the more it is exposed to radiation. In this way, radiation exposure can be carefully monitored.

Other physical barriers are used to protect people from ionising radiation:
- Radiographers use lead screens to prevent exposure.
- Nuclear reactors are encased in thick lead and concrete to prevent radiation escaping into the environment.
- Nuclear technicians going into areas of high levels of radiation must wear a radiation suit made from materials that act as a shield against the radiation.

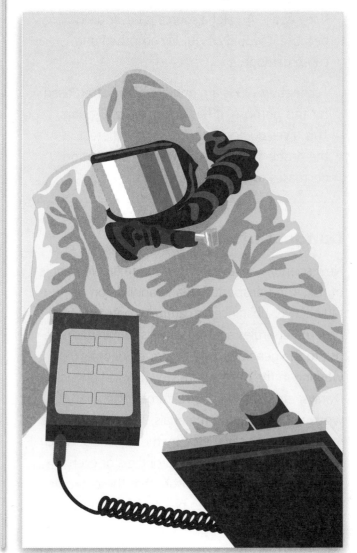

Nuclear Fission

In a chemical reaction it is the electrons that bring about the change. The elements involved remain the same but join up in different ways.

A fission reaction takes place in the nucleus of the atom and different elements are formed. A neutron is absorbed by a large and unstable **uranium** or **plutonium** nucleus. This splits the nucleus into two, roughly equal-sized, smaller nuclei and releases energy and more neutrons.

A fission reaction releases far more energy than that released from a chemical reaction involving a similar mass of material. Once fission has taken place the neutrons released can be absorbed by other nuclei and further fission reactions can take place. This is called a **chain reaction.**

A chain reaction occurs when there is enough fissile material to prevent too many neutrons escaping without being absorbed. This is called critical mass and ensures every reaction triggers at least one further reaction (see opposite).

The Nuclear Reactor

Nuclear power stations use fission reactions to generate the heat needed to boil water into steam. The reactor controls the **chain reaction** so that the energy is released at a steady rate.

Fission occurs in the **fuel rods** and causes them to become very hot.

The **coolant** is a liquid that is pumped through the reactor. The coolant heats up and is then used in the heat exchanger to turn water into steam.

Control rods, made of boron, absorb neutrons, preventing the chain reaction getting out of control. Moving the control rods in and out of the reactor core changes the amount of fission that takes place.

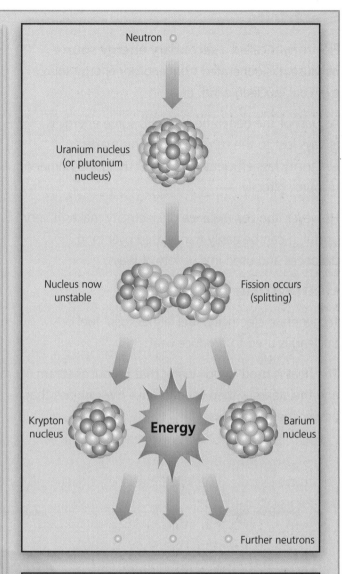

Neutron

Uranium nucleus (or plutonium nucleus)

Nucleus now unstable — Fission occurs (splitting)

Krypton nucleus — **Energy** — Barium nucleus

Further neutrons

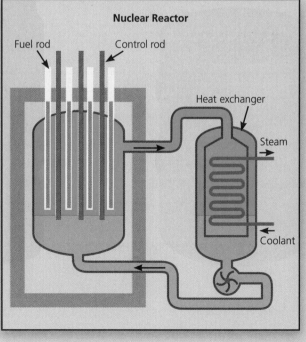

Nuclear Reactor

Fuel rod — Control rod

Heat exchanger

Steam

Coolant

Electricity

Electricity is called a **secondary energy source** because it is generated from another energy source e.g. coal, nuclear, wind, etc.

As part of the generation process some energy is always lost to the surroundings. This makes electricity less efficient than when using the primary resource directly.

However, the convenience of electricity makes it very useful. It can be easily transmitted over long distances and used in a variety of ways.

Generating Electricity

To generate electricity, fuel (either fossil fuel or nuclear) is used to produce heat.

The heat is used to boil water that produces steam, and the steam is then used to drive the turbines that power the generators.

The electricity produced in the generators is sent to a transformer and then to the National Grid, from where we can access it in our homes.

In a **nuclear** power station, the energy is released due to changes in the nucleus of radioactive substances. Nuclear power stations do not produce carbon dioxide but they do produce **radioactive waste**. This nuclear waste is categorised into three types:

- **High-level waste** (HLW) – very radioactive waste that has to be stored carefully. Fortunately, only small amounts are produced and its activity decreases quickly, so it is put into short-term storage.
- **Intermediate-level waste** (ILW) – not as radioactive as HLW but it does remain radioactive for thousands of years. The amount produced is increasing; deciding how to store it safely and permanently is a problem. At the moment most ILW is mixed with concrete and stored in big containers, but this is not a permanent solution.
- **Low-level waste** (LLW) – only slightly radioactive waste that is sealed and placed in landfills.

Low-level Waste

Sealed in steel drums

Intermediate-level Waste

Sealed in concrete, surrounded by steel

High-level Waste

Sealed in glass, surrounded by steel

P6 **1** Which of the following statements about ionising radiation are true? Put ticks (✓) in the boxes next to the **two** correct answers. **[2]**

A Ionising radiation can break molecules into bits called ions. ☐

B X-rays are ionising radiation; gamma rays are not. ☐

C Ionising radiation cannot break molecules into bits called ions. ☐

D X-rays and gamma rays are ionising radiations. ☐

P4 **2** **(a)** A car has a mass of 800kg and is travelling at 20m/s. Calculate the momentum of the car. Include units in your answer. **[3]**

(b) How will the momentum change if the car is loaded with bricks but still travels at 20m/s? **[1]**

(c) How will the momentum change if the car speeds up? **[1]**

P5 **3** **(a)** The following circuit has two identical bulbs. Write down the missing values for:

(i) A_1 **[1]**

(ii) A_3 **[1]**

(iii) V_2 **[1]**

(iv) V_3 **[1]**

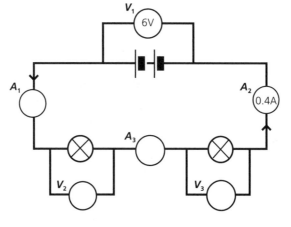

(b) Draw straight lines between the boxes to join the circuit symbols with their correct meaning. **[3]**

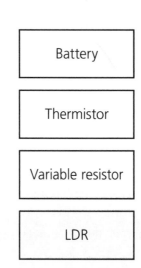

	Battery
	Thermistor
	Variable resistor
	LDR

Exam Practice Questions

P6 **4** Name three sources that contribute to background radiation. **[3]**

P4 **5** A school bus sets off from school and drives to its first stop. Describe the motion of the bus, using the distance–time graph below. **[3]**

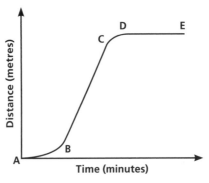

P6 **6** This question is about nuclear power.

I'm the manager of the Super Fuels Nuclear Power Company. Nuclear energy is free energy and it doesn't produce any pollution.

Mr Benbow

I live opposite the power station, and I think nuclear power is dangerous, like a ticking time bomb.

Miss Manning

I work for the electricity board. I think that nuclear power is the only alternative to fossil fuels.

Mr Holmes

(a) For two of the above people, explain why their point of view may be unreliable. **[1]**

(b) For one of the above people, explain why their point of view may be wrong. **[1]**

P6 **7** In 1911, Rutherford and Marsden carried out an experiment that disproved the 'plum pudding' theory that described the structure of the atom. They fired alpha particles at gold foil in a vacuum and recorded the paths of the alpha particles.

Describe the observations made and the conclusions drawn from the experiment. **[6]**

✎ *The quality of written communication will be assessed in your answer to this question.*

HT **P6** **8** Potassium-40 decays by beta emission to argon-40. A sample of Moon rock that originally contained no argon-40 atoms now contains 15 atoms of potassium-40 for every atom of argon-40. The half-life of potassium-40 is 1 billion years. How old is the Moon rock? **[2]**

Modules B4–B6 (Pages 30–31)

1.

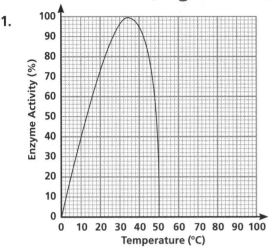

[1 mark for a curve drawn with a single maximum; 1 mark for the peak of the curve at 35°C; 1 mark for a rapid decrease in enzyme activity after 40°C]

2. **This is a model answer which would score full marks:** Aerobic respiration involves the reactants glucose and oxygen reacting to produce water and carbon dioxide, with the release of energy. Anaerobic respiration in humans involves glucose being converted to lactic acid, with the release of energy. In plants and microorganisms such as yeast the products of anaerobic respiration are ethanol and carbon dioxide, with energy released. The energy produced by aerobic respiration is much greater than that produced by anaerobic respiration.

3. Water moving from plant cell to plant cell; A pear losing water in a concentrated salt solution; **and** Water moving from blood to body cells **should be ticked**.

4. have a scientific mechanism; be published; **and** be repeatable **should be ticked [2 marks for all three; 1 mark for two]**.

5. 16 cell stage / After the 8 cell stage

6. **(a)** The plant will grow to the left / towards the light.

(b) survival; nervous system / CNS

7. **(a)** **Any three from:** Stepping; Grasping; Startle; Sucking; Rooting

(b) **This is a model answer which would score full marks:** Reflexes enable an organism to react to a stimulus without thinking. The response is quicker than it would be if the organism were to think about how to respond. This may mean the organism survives longer. For example, some simple animals react to changes in light levels, which could indicate a predator was overhead.

(c) Conditioned reflexes

8. $C_6H_{12}O_6 + 6O_2 \rightarrow 6CO_2 + 6H_2O$
[1 mark for reactants; 1 mark for products]

9. UUAUCGGCGGUUAAUCCG

10. Ryan

Answers

Modules C4–C6 (Pages 66–67)

1. (a) (i) H_2O (ii) CH_4

    (b) C + 2 O = 12 + (2 × 16) = 44 **[1 mark for correct working but wrong answer]**

2. (a) Reduction **should be ringed.**

    (b) (i) Re-use; Recycle; Throw away / landfill
       (ii) **Any suitable answer, e.g.** It is best to re-use where possible as this has no environmental impact **[1]**. Throwing things away has the most impact as they will go to landfill **[1]**. Landfill sites do not look good and they remove natural habitats **[1]**. Recycling is better than throwing away as it uses a lot less energy than the initial manufacturing **[1]**.

3. (a) (i) A light blue precipitate
       (ii) A white precipitate

    (b) Iron(III) iodide **[1]**. A red-brown precipitate forming with sodium hydroxide is the positive test for iron(III) **[1]**. A yellow precipitate forming with acidified silver nitrate is the positive test for iodide ions **[1]**.

4. (a) The reaction is complete **[1]** and so no more product can be made **[1]**.

    (b) Reaction A was at a higher temperature **[1]**, had smaller particles or a larger surface area **[1]**, or there was a catalyst present **[1]**.

    (c) Neither reaction **[1]**. They both finished with the same amount of product as the horizontal line is in the same place **[1]**.

5. (a) Sodium reacts with cold water **and** Sodium tarnishes in air to make sodium oxide **should be ticked.**

    (b) (i) 11 protons; 12 neutrons
       (ii) **The diagram should have two crosses on the first shell, eight on the second shell and one on the third shell.**
       (iii) 2.8.1

    (c) (i) **Any two from:** Work in a fume cupboard; Wear safety goggles; Wear protective clothing, e.g. gloves, coat; Use a safety screen.
       (ii) $2Na + Cl_2 \longrightarrow 2NaCl$ **[1 mark for correct symbols; 1 mark for 2Na]**
       (iii) **In this order:** (s); (g); (s)

    (d) **This is a model answer which would score full marks:** The electron in the outer shell in the sodium atom will go into the outer shell of the chlorine atom. This will leave the sodium atom with a positive charge and the chlorine atom with a negative charge. The sodium and chloride ions are attracted together to form a crystal lattice of sodium chloride.

    (e) The reactivity of alkali metals increases as you go down the group **[1]**. This is because the outermost electron is further away from the influence of the nucleus and so an electron is lost more easily **[1]**. Potassium is below sodium in Group 1, so the reaction between potassium and chlorine is faster **[1]**.

Answers

Modules P4–P6 (Pages 97–98)

1. Ionising radiation can break molecules into bits called ions **and** X-rays and gamma rays are ionising radiation **should be ticked**.

2. (a) Momentum = 800kg × 20m/s = 16 000kg m/s
   [2 marks for correct answer; 1 mark for correct units]

   (b) The momentum will increase.

   (c) The momentum will increase.

3. (a) (i) 0.4A (ii) 0.4A (iii) 3V (iv) 3V

   (b)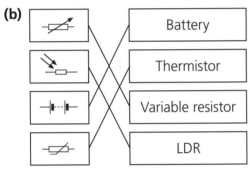

	Battery
	Thermistor
	Variable resistor
	LDR

   [1 mark for each correct line up to a maximum of 3.]

4. **Any three from:** Medical industry; Nuclear industry; Cosmic rays; X-rays; Food; Radon gas; Stars; Rocks; Uranium; Nuclear weapons testing

5. Between A and B, the bus accelerates from a stationary position **[1]**. The bus then travels at a constant speed to C **[1]**. The bus decelerates between C and D, and is stationary between D and E **[1]**.

6. (a) Mr Benbow and Mr Holmes work for companies that depend on nuclear power.

   (b) **Any suitable answer, e.g.**
   Mr Benbow – Nuclear energy is not free as the cost of commissioning and decommissioning nuclear plants is very high. Pollution does exist in the form of radioactive waste.
   Miss Manning – The nuclear industry has a better safety record than fossil-fuel power stations.
   Mr Holmes – Renewable energy resources have not been invested in as much as they could. For example, the UK could use wind power to supply a much greater proportion of the electricity needed than it currently does.

7. **This is a model answer which would score full marks:** The vast majority of alpha particles passed through the gold foil without being deflected. Rutherford and Marsden concluded that the atom was mostly empty space. Some alpha particles were deflected through a small angle. Alpha particles are positively charged and they were being repelled by another positively charged object. A very small number of alpha particles were deflected through a large angle (greater than 90°) and for this to happen there must have been a concentration of positive charge in a very small area, i.e. a nucleus of positive charge.

8. 16 → 8 → 4 → 2 → 1 so 4 half-lives have occurred; 4 × 1 billion = 4 billion years
   [If answer is incorrect, 1 mark for calculation based on half-life]

Glossary

Acceleration – how quickly speed increases or decreases (per second squared).

Acid – an aqueous compound with a pH value less than 7.

Activation energy – the minimum amount of energy required to cause a reaction.

Activity – the amount of radioactive decay per second.

Aerobic respiration – respiration using oxygen; releases energy and produces carbon dioxide and water.

Alkali – a substance that has a pH value higher than 7.

Alternating current – a current that reverses direction of flow continuously.

Anaerobic respiration – respiration that takes place in the absence of oxygen.

Aqueous solution – a solution made when a solute dissolves in water.

Atmosphere – the layer of gases surrounding the Earth.

Atom – the smallest part of an element that can enter into a chemical reaction.

Attraction – the force experienced by two objects with mass; the force between two objects with opposite electric charge.

Biotechnology – the use of organisms, parts of organisms, or the process the organisms carry out, to produce useful (chemical) substances.

Bulk chemicals – chemicals made on a large scale.

Catalyst – a substance that is used to speed up a chemical reaction without being chemically altered itself.

Cell – the fundamental unit of a living organism; a component that produces a voltage in a circuit.

Central nervous system (CNS) – the brain and spinal cord; allows an organism to react to its surroundings and coordinates its responses.

Chemical synthesis – the process by which many useful products are made.

Chromosome – a coil of DNA made up of genes, found in the nucleus of plant / animal cells.

Clone – a genetically identical offspring of an organism.

Combustion – a chemical reaction that occurs when fuels burn, releasing heat.

Compound – a substance in which the atoms of two or more elements are chemically joined, either by ionic or covalent bonds.

Conditioned reflex – a specific reflex response produced when a certain stimulus is detected.

Covalent bond – the force of attraction between two atoms sharing electrons.

Crystallisation – the formation of solid crystals from a solution.

Cytoplasm – the protoplasm of the cell which is found outside the nucleus.

Diffusion – the movement of particles from high to low concentration down a concentration gradient.

Direct current – an electric current that only flows in one direction.

Distance–time graph – a graph showing distance travelled against time taken; the gradient of the line represents speed.

DNA (deoxyribonucleic acid) – the nucleic acid molecules that make up chromosomes and carry genetic information; found in every cell of every organism; control cell chemistry.

Effector – the part of the body, e.g. a muscle or a gland, which produces a response to a stimulus.

Electrolysis – the process by which an electric current causes a solution, containing ions, to undergo chemical decomposition.

Electron – a negatively charged particle that orbits the nucleus.

Element – a substance that consists of one type of atom.

Embryo – a ball of cells that will develop into a human/animal baby.

Endothermic reaction – a chemical reaction that takes in energy (heat) from its surroundings so that the products have more energy than the reactants.

Enzyme – a protein molecule and biological catalyst found in living organisms that helps chemical reactions to take place (usually by increasing the rate of reaction).

Evaporation – the process in which a liquid changes into a gas by heating.

Exothermic reaction – a chemical reaction that gives out energy (heat) to its surroundings so that the products have less energy than the reactants.

Fertilisation – the fusion of a male gamete with a female gamete.

Filtration – a method for separating solids from liquids by passing a mixture through a porous material.

Fine chemicals – chemicals manufactured on a small scale.

Food chain – a representation of the feeding relationship between organisms; energy is transferred up the chain.

Force – a push or pull acting upon an object.

Frequency – the number of times that something happens in a set period of time; the number of times a wave oscillates in one second; measured in hertz.

Fuel – a substance that releases energy when burned in the presence of oxygen.

Gamete – a specialised sex cell.

Gene – a small section of DNA, in a chromosome, that determines a particular characteristic; controls cellular activity by providing instructions (coding) for the production of a specific protein.

Gravitational potential energy – the energy an object has because of its height and mass above the Earth.

Group – a vertical column of elements in the periodic table.

Half-life – the time taken for half the radioactive nuclei in a material to decay.

Halogen – an element found in Group 7 of the periodic table.

Heart – a muscular organ that pumps blood around the body.

Hormone – a chemical messenger, made in ductless glands, which travels around the body in the blood to affect target organs elsewhere in the body.

Hydrosphere – contains all the water on Earth, including rivers, oceans, lakes, etc.

Insoluble – a property that means a substance cannot dissolve in a solvent.

Instantaneous speed – the speed of an object at a particular moment.

Ion – a positively or negatively charged particle formed when an atom, or groups of atoms, loses or gains electrons.

Irradiation – exposure to ionising radiation.

IVF (*In vitro* fertilisation) – a technique in which egg cells are fertilised outside the woman's body.

Kinetic energy – the energy possessed by an object because of its movement.

Life cycle assessment (LCA) – an analysis of a product from manufacture to disposal.

Lithosphere – the rigid outer layer of the Earth made up of the crust and the part of the mantle just below it.

Glossary

Meiosis – the type of cell division that forms daughter cells with half the number of chromosomes as the parent cell; produces gametes.

Meristem – an area where unspecialised cells divide, producing plant growth, e.g. roots and shoots.

Mitochondria – membrane-bound organelles containing enzymes that carry out respiration for a cell.

Mitosis – the type of cell division that forms two daughter cells, each with the same number of chromosomes as the parent cell.

Momentum – a measure of the state of motion of an object as a product of its mass and velocity.

Muscle – tissue that can contract and relax to produce movement.

Neuron – a specialised cell that transmits electrical messages (nerve impulses) when stimulated.

Neutralisation – the reaction between an acid and a base which forms products that are pH neutral.

Neutron – a particle found in the nucleus of an atom that has no electric charge.

Nuclear fusion – light nuclei join to form a heavier nucleus, releasing energy.

Organelle – a specialised subunit of a cell with its own function.

Osmosis – the net movement of water from a dilute solution (lots of water) to a more concentrated solution (little water) across a partially permeable membrane.

Oxidation – a chemical reaction that occurs when oxygen joins with an element or compound.

Parallel circuit – a circuit in which components are connected in separate loops, so there are two (or more) paths for the current to take.

Period – a horizontal row of elements in the periodic table.

pH – a measure of acidity or alkalinity.

Photosynthesis – the chemical process that takes place in green plants where water combines with carbon dioxide to produce glucose using light energy.

Phototropism – a plant's response to light.

Plasmid – a section of DNA in a bacterium that reproduces independently of the main chromosome.

Pollutant – a chemical that can harm the environment and health.

Potential difference (voltage) – the difference in electrical energy carried by the charge between two points.

Precipitate – an insoluble solid formed during a reaction involving solutions.

Precipitation – the process of forming a precipitate by mixing solutions.

Proton – a positively charged particle found in the nucleus of an atom.

Receptor – the part of the nervous system that detects a stimulus; a sense organ, e.g. eyes, ears, nose, etc.

Reduction – a chemical reaction that occurs when oxygen is removed.

Reflex action – an involuntary action, e.g. automatically jerking your hand away from something hot; a fast, involuntary response to a stimulus.

Relative atomic mass (RAM or A_r) – the average mass of an atom of an element compared with a twelfth of the mass of a carbon atom.

Relative formula mass (RFM or M_r) – the sum of the atomic masses of all the atoms in a molecule.

Repulsion – a term used to describe the effect when two materials with the same charge push away from (repel) each other.

Resistance – a measure of how hard it is to get a current through a component at a particular potential difference.

Respiration – the liberation of energy from food.

Resultant force – the total force acting on an object (the effect of all the forces combined).

Ribosome – a small structure found in the cytoplasm of living cells, where protein synthesis takes place.

Salt – the product of a chemical reaction between a base and an acid.

Series circuit – a circuit in which there is only one path for the current to take; all components are connected in a single loop.

Sexual reproduction – reproduction involving the fusing together of gametes formed via meiosis.

Soluble – a property that means a substance can dissolve in a solvent.

Solution – the mixture formed when a solute dissolves in a solvent.

Solvent – a liquid that can dissolve another substance to produce a solution.

Specialised – developed or adapted for a specific function.

Species – a group of organisms capable of breeding to produce fertile offspring.

Speed – how far an object travels in a given time.

Static electricity – the transfer of electrons from one material to another.

Stem cell – a cell that can give rise to specialised cells.

Stimulus – a change in the environment that can elicit a response in a living organism.

Synapse – the small gap between adjacent neurons.

Titration – an accurate technique that can be used to find the volume of liquid needed to neutralise an acid.

Transformer – an electrical device used to change the voltage of alternating currents.

Universal indicator – a mixture of pH indicators, which produces a range of colours according to pH and can, therefore, be used to measure the pH of a solution.

Velocity – an object's speed and direction.

Voltage (potential difference) – the difference in electrical energy carried by the charge between two points.

Yield – the amount of product obtained, e.g. from a crop or a chemical reaction.

Zygote – the first cell formed after the fertilisation of an egg by a sperm.

HT

Active transport – the movement of substances against a concentration gradient; requires energy.

Auxin – a plant hormone that affects the growth and development of the plant.

Chain reaction – a reaction, for example nuclear fission, that is self-sustaining.

Denatured – the state of an enzyme that has been destroyed by heat or pH and can no longer work.

Isotope – an atom of the same element that contains different numbers of neutrons but the same number of protons.

Nuclear fission – the splitting of atomic nuclei, which is accompanied by a release in energy.

Nuclear reactor – the place where fission occurs in a nuclear power station.

Velocity–time graph – a graph showing velocity against time taken; the line represents acceleration.

Periodic Table

Key

| relative atomic mass |
| **atomic symbol** |
| name |
| atomic (proton) number |

1
H
hydrogen
1

Group 1	Group 2											Group 3	Group 4	Group 5	Group 6	Group 7	Group 0
																	4 **He** helium 2
7 **Li** lithium 3	9 **Be** beryllium 4											11 **B** boron 5	12 **C** carbon 6	14 **N** nitrogen 7	16 **O** oxygen 8	19 **F** fluorine 9	20 **Ne** neon 10
23 **Na** sodium 11	24 **Mg** magnesium 12											27 **Al** aluminium 13	28 **Si** silicon 14	31 **P** phosphorus 15	32 **S** sulfur 16	35.5 **Cl** chlorine 17	40 **Ar** argon 18
39 **K** potassium 19	40 **Ca** calcium 20	45 **Sc** scandium 21	48 **Ti** titanium 22	51 **V** vanadium 23	52 **Cr** chromium 24	55 **Mn** manganese 25	56 **Fe** iron 26	59 **Co** cobalt 27	59 **Ni** nickel 28	63.5 **Cu** copper 29	65 **Zn** zinc 30	70 **Ga** gallium 31	73 **Ge** germanium 32	75 **As** arsenic 33	79 **Se** selenium 34	80 **Br** bromine 35	84 **Kr** krypton 36
85 **Rb** rubidium 37	88 **Sr** strontium 38	89 **Y** yttrium 39	91 **Zr** zirconium 40	93 **Nb** niobium 41	96 **Mo** molybdenum 42	[98] **Tc** technetium 43	101 **Ru** ruthenium 44	103 **Rh** rhodium 45	106 **Pd** palladium 46	108 **Ag** silver 47	112 **Cd** cadmium 48	115 **In** indium 49	119 **Sn** tin 50	122 **Sb** antimony 51	128 **Te** tellurium 52	127 **I** iodine 53	131 **Xe** xenon 54
133 **Cs** caesium 55	137 **Ba** barium 56	139 **La*** lanthanum 57	178 **Hf** hafnium 72	181 **Ta** tantalum 73	184 **W** tungsten 74	186 **Re** rhenium 75	190 **Os** osmium 76	192 **Ir** iridium 77	195 **Pt** platinum 78	197 **Au** gold 79	201 **Hg** mercury 80	204 **Tl** thallium 81	207 **Pb** lead 82	209 **Bi** bismuth 83	[209] **Po** polonium 84	[210] **At** astatine 85	[222] **Rn** radon 86
[223] **Fr** francium 87	[226] **Ra** radium 88	[227] **Ac*** actinium 89	[261] **Rf** rutherfordium 104	[262] **Db** dubnium 105	[266] **Sg** seaborgium 106	[264] **Bh** bohrium 107	[277] **Hs** hassium 108	[268] **Mt** meitnerium 109	[271] **Ds** darmstadtium 110	[272] **Rg** roentgenium 111							

Elements with atomic numbers 112–116 have been reported but not fully authenticated

The lanthanoids (atomic numbers 58–71) and the actinoids (atomic numbers (90–103) have been omitted.

The relative atomic masses of copper and chlorine have not been rounded to the nearest whole number.

Qualitative Analysis

Tests for Positively Charged Ions

Ion	Test	Observation
Calcium Ca^{2+}	Add dilute sodium hydroxide	A white precipitate forms; the precipitate does not dissolve in excess sodium hydroxide
Copper Cu^{2+}	Add dilute sodium hydroxide	A light blue precipitate forms; the precipitate does not dissolve in excess sodium hydroxide
Iron(II) Fe^{2+}	Add dilute sodium hydroxide	A green precipitate forms; the precipitate does not dissolve in excess sodium hydroxide
Iron(III) Fe^{3+}	Add dilute sodium hydroxide	A red-brown precipitate forms; the precipitate does not dissolve in excess sodium hydroxide
Zinc Zn^{2+}	Add dilute sodium hydroxide	A white precipitate forms; the precipitate dissolves in excess sodium hydroxide

Tests for Negatively Charged Ions

Ion	Test	Observation
Carbonate CO_3^{2-}	Add dilute acid	The solution effervesces; carbon dioxide gas is produced (the gas turns limewater from colourless to milky)
Chloride Cl^-	Add dilute nitric acid, then add silver nitrate	A white precipitate forms
Bromide Br^-	Add dilute nitric acid, then add silver nitrate	A cream precipitate forms
Iodide I^-	Add dilute nitric acid, then add silver nitrate	A yellow precipitate forms
Sulfate SO_4^{2-}	Add dilute nitric acid, then add barium chloride or barium nitrate	A white precipitate forms

Index